The Walter Lynwood Fleming
Lectures in Southern History

Louisiana State University

Olive Branch
and Sword—
The Compromise
of 1833

Olive Branch and Sword— The Compromise of 1833

MERRILL D. PETERSON

LOUISIANA STATE UNIVERSITY PRESS

BATON ROUGE AND LONDON

1002134 807

Copyright © 1982 by Louisiana State University Press
All rights reserved
Manufactured in the United States of America

Designer: Albert Crochet
Typeface: VIP Palatino
Typesetter: G&S Typesetters, Inc.
Printer and Binder: Thomson-Shore, Inc.

LIBRARY OF CONGRESS CATALOGING IN PUBLICATION DATA

Peterson, Merrill D.
   Olive branch and sword.

   (Walter Lynwood Fleming Lectures in southern history)
   Includes index.
   1. Nullification—Addresses, essays, lectures.
I. Title.   II. Series.
E384.3.P47   973.5'61   81-13739
ISBN 0-8071-0894-4   AACR2

T

To Dumas Malone

Contents

Acknowledgments

This book is based upon the Walter Lynwood Fleming Lectures I delivered at Louisiana State University in 1980. It is a by-product of a larger work, *The Great Triumvirate: Webster, Clay, and Calhoun*, that I hope to complete during the next several years. As a glance at the notes will suggest, I am indebted to many libraries and manuscript depositories, from Boston to Baton Rouge; and I wish particularly to thank the editors of The Papers of John C. Calhoun and The Papers of Henry Clay for their assistance. My research was aided by grants from the American Philosophical Society and the American Council of Learned Societies. I am pleased to acknowledge their support. Richard E. Ellis read the manuscript for Louisiana State University Press. I am grateful to him for his criticism and apologize for not responding to it more fully. Finally, I wish to thank Professor John Loos and his colleagues at Louisiana State University for the honor of appearing on such a distinguished lecture platform, for their attentiveness, and for their hospitality.

I Matrix

On March 2, 1833, amidst the deluge of legislation passed in the closing hours of the Twenty-second Congress, President Andrew Jackson signed into law two bills that ended the crisis of the Union provoked by South Carolina's nullification of the protective tariff just over three months before. The Compromise Tariff went some way toward meeting South Carolina's objections to protectionism by scheduling a gradual reduction of duties to the revenue level of 20 percent over a period of nine and one-half years. Although protectionist champion Henry Clay was its author, the chief of the Nullifiers, John C. Calhoun, nevertheless embraced it, and thus it had the character of a treaty of peace between opposing parties, policies, and interests. The other act, the Force Bill as it was called, authorized the president to employ the military, if necessary, to enforce the revenue laws and put down nullification in South Carolina. The two acts were, respectively, the olive branch and the sword of the great compromise that preserved the peace, the Constitution, and the Union in 1833.

The history of the American republic from the Revolution to the Civil War is punctuated by chapters of crisis and compromise. The Union itself was a compromise, like the Constitution, whose equivocal character—"partly national, partly federal" in the Madisonian formulation—seemed to require ongoing accommodation and compromise as the price of survival. The task of the second generation of American statesmen was markedly different from that of the founders—not of making a union and a constitution but of preserving them, not of securing liberty and self-government but of embellishing them. In a rapidly expanding and modernizing society the fulfillment of "this sacred trust" was beset with dangers

1

both real and imaginary. The art of compromise became a conserving force, cushioning the shocks of change, mediating between reason and violence, vindicating institutions at the expense of interests or doctrines; and when this art failed, finally, the Union was dissolved. The story is familiar. Historians of the middle period have written exhaustively of the Missouri Compromise of 1820 and the Compromise of 1850. But they have said very little about the Compromise of 1833, midway between these two great landmarks. It has generally been treated, wearily, as the tail end of the nullification controversy; for some more recent students it has been obscured entirely by the pyrotechnics of the Jacksonian Bank War. The Compromise of 1833 merits attention, however, not only for its importance in the nation's political history but as a case study in the politics of compromise itself.

Compromise, by definition, involves the mutual concession of principle or policy or interest, or all together, between conflicting parties who discover they have more to gain than to lose by concessions. "The essence of politics is compromise," as Lord Macaulay remarked. But great compromise, as in the historic American Union, transcended ordinary politics and invites more particular analysis. Who were the parties to the Compromise of 1833? Who were its architects? Who were its beneficiaries? To these questions, as simple as they may seem, there are no simple answers.

What was being conceded by the two sides? The existing Tariff of 1832, prospectively, and the South Carolina ordinance nullifying it, immediately; but these were only the legislative expressions of a conflict of ideas and interests that were much too powerful to be contained by the exigencies of compromise.

What were the motives of the makers of compromise? A fair share of patriotism and high purpose could not be denied to them, but those who traded principle for conciliation in a fiercely partisan political climate were inevitably suspect. Leaders like Jackson, Clay, and Calhoun were more than ordinary politicians. They were public symbols of ideas, personifications of policies, heroes of political battles that altered the course of the nation; and it is impossible to understand the events surrounding the compromise without understanding them.

What was pledged as to the future? No mere act of legislation could bind future congresses, though a moral obligation might be assumed, and ambiguities and contingencies written into the Compromise Act seemed to ensure it would be pushed and pulled in different directions. There was, indeed, disagreement from the beginning even on the basic elements of the compromise. When President Jackson signed the tariff and force bills into law, he also pointedly refused to sign, and subsequently vetoed, still another bill providing for distribution of the proceeds from the public lands among the states; yet some congressmen considered it the indispensable third leg of the compromise. This would be a source of recrimination in the years to come.

The symbols of olive branch and sword, while expressive of the spirit of the compromise, cannot explain it. Setting out to explain it, I will first look at the dynamics of policy and politics that led to the national crisis of 1832–1833, then inquire closely into the making of the compromise, and conclude with an examination of the consequences and effects of the compromise during the dozen years that it remained a public issue.

In the years following the Peace of Ghent, as the nation turned away from Europe inward to the development of its own resources, a system of political economy emerged that Henry Clay, its leading congressional advocate, named the American System. It sought to promote growth and productivity; it assigned the national government a positive role in economic development; it was centralizing in policy and effect. In this character the American System was sometimes mistaken for a renascent Hamiltonian Federalism, alerting old fears for the survival of the republic. In fact, whatever its policy debt to this tradition, the American System was the legitimate outcome of Jeffersonian experience at the helm of government since 1801. It was part of the Jeffersonian revulsion from Europe, from an economic prosperity floated on the perils of the

Atlantic carrying trade, at the terrible cost, finally, of war. At home, as Jefferson had predicted, the Republicans had *become* the nation; and so they could be inspired more by their hopes than by their fears of governmental power. Uplifted by the patriotic élan of 1815, confident that the republic was secure at last, nearly all Republicans could unite on policies of national improvement and consolidation. The consensus formed around the program President James Madison sent to Congress in that year. This "Madisonian platform" was the basis of the American System.[1]

The central measure, the *primum mobile*, of the system was the protective tariff. Although the protection and encouragement of domestic manufactures was an object of the government's earliest tariff laws, it became the primary object, overriding the concern for revenue, only after the War of 1812. A "new epoch" had dawned, as Clay declared. During most of its history the United States had shaped its industry and commerce to foreign markets thrown open by the convulsions of war and revolution in Europe—markets destined to disappear with the return of peace. Embargo and war on the American side, meanwhile, had spurred the growth of domestic manufactures. To protect these still-struggling infants from the hazards of foreign competition, to give employment to American labor, to create domestic markets for the country's agricultural surplus, became accepted goals of national policy.

Advocates of protection built an elaborate ideology. It was freighted with the old rhetoric of American independence, for, the argument ran, the country was still vulnerable economically to Great Britain, who in her jealousy of "the youthful Hercules" sought to strangle American manufactures in the cradle. The ideology rejected the laissez-faire and free-market maxims of the new school of economic theorists, insisting that these maxims while everywhere proclaimed were nowhere practiced and that to impose them on a young and developing country would be to condemn it to permanent poverty, weak-

1. Madison's Seventh Annual Message is in J. D. Richardson (ed.), *A Compilation of the Messages and Papers of the Presidents* (Washington, D.C., 1907), I, 547–54.

ness, and inferiority. A conception of national wealth took hold; it was an aggregate interest, paramount to the interests of individuals or of other nations, and in the United States the possibilities of its development were wonderful to behold. The first generation of American republicans had worked out the implications of a large territory for free government under the Constitution and found them good. Now statesmen and philosophers of the second generation worked out the implications for the American economy, with the same positive result.

A country of continental proportions, a vast and wonderfully varied landed empire with almost limitless room to grow, was potentially "a world within itself." It need not, like Britain, look to far-flung markets that might be conquered under a regime of free trade; it could look instead to the development of a "territorial division of labor," founded in climate and geography, and to a protected "home market" for the productions of its diverse industry. In the favorite model fashioned by Clay and his associates, each great section of the country would concentrate on the productions for which it was best suited: the South on staples like cotton, the West on grains and livestock, the Northeast on manufacturing. A mutually supporting and balanced economy of agriculture, manufactures, and commerce would thus be established; and although founded on sectional interests, the sum of the whole, the *national* interest, would be greater than the sum of the parts.[2]

An expanding home market called for improved transportation facilities. Internal improvements, therefore, was a leading measure of the American System. National defense, the rapid growth of the Transappalachian West, and the need to

2. See the speeches of Henry Clay in the House of Representatives, April 26, 1820, and March 30–31, 1824, in Calvin Colton (ed.), *The Works of Henry Clay* (New York, 1904), VI, 219–37, 255–94. Among the important expositions of protectionism: Mathew Carey, *Addresses of the Philadelphia Society for the Promotion of National Industry* (Philadelphia, 1819); Friedrich List, *Outlines of American Political Economy* (Philadelphia, 1827); and the constant stream of editorial writings in *Niles' Weekly Register*, 1815–1833.

bind the Union together offered ample reasons, quite aside
from the economic ones, for a comprehensive system of roads
and canals. "Let us conquer space," Calhoun imperiously declared in 1817. He was impatient with Republicans (and a few
old Federalists) who raised constitutional objections to a national system. "It must be submitted to as a condition of our
greatness," he said.[3] How great, how harmonious, how felici-
tous the new order might become is suggested by the report
of a congressional committee in 1816.

> Different sections of the union will according to their position, the
> climate, the population, the habits of the people, and the nature
> of the soil, strike into that line of industry which is best adapted to
> their interest and the good of the whole; an active and free inter-
> course, promoted and facilitated by roads and canals, will ensue;
> prejudices which are generated by distance . . . will be removed;
> information will be extended, the union will acquire strength and
> solidarity, and the Constitution of the United States, and that of
> each state, will be regarded as fountains from which flow numer-
> ous streams of public and private prosperity.[4]

Internal improvements would be financed in part, especially
after retirement of the national debt, by revenue from the sale
of the public lands. The lands were a great resource, of course,
and no one advocated retreat from the liberal policy for disposing of them; but they were not an unmixed blessing in the view
of economic nationalists. The wild lands steadily drained off
labor and capital from the East, where they were wanted for
the growth of manufactures and the whole complex of arts and
industries of advanced civilization. One of the benefits of the
protective tariff, some said, was that it helped to check this
potentially crippling diffusion of national energies. It could
not be fairly criticized on grounds of equity, for the support
it extended to manufacturing was more than offset by the

---

3. Speech on the Bonus Bill, February 4, 1817, in Robert L. Meriwether
(ed.), *The Papers of John C. Calhoun* (Columbia, S.C., 1859), I, 401. On internal
improvements, see Carter Goodrich, "National Planning of Internal Improvements," *Political Science Quarterly*, LXIII (1948), 16–44.
4. Report of the Committee on Commerce and Manufactures, House of
Representatives, February 13, 1816, in *Niles' Weekly Register*, February 24, 1816.

bounty of public lands policy to western agriculture.[5] Still another element of the American System, though controversial even among leading proponents, was a national bank to facilitate credit expansion, manage domestic exchange, and secure a stable and uniform currency throughout the country. The Second Bank of the United States, chartered in 1816, gradually took on these responsibilities. Still other and lesser measures, such as a national bankruptcy law, never found adequate support.

Unfortunately for the American System, the promising consensus of the postwar years broke down in the 1820s, and it was politically impossible thereafter to restore the ideal of a government positively charged with the promotion of the national interest. Widespread economic distress followed in the wake of the Panic of 1819. In the East, where the panic bankrupted merchants and manufacturers, idled workers, and closed factories, the cry went up for higher tariffs. But the panic struck hardest in the farmlands of the Ohio Valley and in the burgeoning cotton lands of the lower South. To the extent that the ensuing depression sold westerners on the idea of the home market, it strengthened their ties to the American System and they joined eastern allies in the movement for higher tariffs. Many westerners, however, traced their misfortunes to the Bank of the United States, which, they believed, triggered the economic collapse by severely contracting credit in order to save itself. "The Bank was saved, the people were ruined," it was commonly said. Anger against the bank was converted into anger against the government that chartered this "monster" and against the East, where its power and profits lay. Anger was turned on the national land system as well, for westerners who had bought lands on easy credit now lost them, and they began to question the justice, and the reputed liberality, of the system. In the South, where staple prices plunged downward, the panic was the signal for retreat from policies of economic nationalism. Southern sup-

5. See especially the annual report of the secretary of treasury, Richard Rush, December 8, 1827, *ibid.*, December 15, 1827.

port for the tariff vanished; and cotton planters took up the
cause of free trade heretofore championed by northern merchants and shippers in overseas commerce. At the same time,
anxieties and fears touching slavery, newly aroused by the
debate over the Missouri Compromise, sharpened southern
sensitivities to the balance of sectional power in the Union. [6]

The remarkable political unity that had won for President
James Monroe's administration the appellation "era of good
feelings" also broke down. No sooner was Monroe inaugu-
rated a second time than the contest for the succession began
among a host of aspirants. Neither the president nor the congressional caucus nor any other instrument of the Republican
party could control it. The party split into personal followings
and factions; great issues of public policy were submitted to
the artifice and caprice of presidential politics. In a parallel development each of the three sections became more and
more conscious of its peculiar economic interest and committed to its pursuit, reducing national politics to a struggle
for power among these great geographical interests. The earlier vision of harmony, balance, and reciprocity among the
sections was shattered. No separate nations ever entertained
more opposing views of public policy, as Calhoun later observed. Meetings of Congress degenerated into "an annual
struggle . . . in which all the noble and generous feelings of
patriotism are gradually subsiding into sectional and selfish
attachments." [7] Since no section could rule by itself, the grand
strategy in national politics was to build a winning coalition of
two against one.

This was still the game in January, 1830, when Missouri
Senator Thomas Hart Benton provoked the celebrated debate
on Foot's Resolution. Calling attention to the disclosure in the

6. Norris W. Preyer, "Southern Support of the Tariff of 1816: A Reapprais-
al," *Journal of Southern History*, XXV (1959), 306–22. On the impact of the Panic
of 1819, see George Dangerfield, *The Era of Good Feelings* (New York, 1953),
Charles S. Sydnor, *The Development of Southern Sectionalism* (Baton Rouge,
1948), and Frederick Jackson Turner, *The Rise of the New West, 1819–1829* (Boston, 1906).

7. Fort Hill Address, July 26, 1831, in Richard K. Crallé (ed.), *Works of John
C. Calhoun* (New York, 1853–55), VI, 78.

report of the commissioner of the Land Office that seventy-two million acres of land already surveyed remained unsold, Samuel A. Foot, of Connecticut, had asked for an inquiry into the wisdom of suspending the surveys. The request was routine, not expected to invite challenge or debate. But Benton, with characteristic bluster, assailed the resolution for proposing to forge still another link in the chain of eastern oppression of the West—a chain reaching back over fifty years in his view. By the control of public land policy, Benton charged, eastern interests sought to check the growth of the West, maintain their political ascendancy, and ensure a pauperized labor force for their mills and factories. The protective tariff was similarly exploitative of the South. Lauding the South as "the ancient defender and savior of the West," Benton begged her again, as in Thomas Jefferson's time, to stretch forth the protecting arm.[8]

There was more than a little awkwardness in this, for regardless of historic ties between the two sections, westerners continued to support leading American System measures, above all internal improvements, while the South was rallying to the old Republican standard of states' rights. Politicians of this faith had never accepted the constitutionality of internal improvements; many who spoke for the South believed the policy survived mainly to keep up the tariff, whose Pactolian riches were drained into great projects for roads and canals. Nevertheless, South Carolina's senior senator, Robert Y. Hayne, responded to Benton's plea. The protective tariff and the public land law were "parallel oppressions," he said, one fatal to the South, the other to the West, and both fatal to the Union. Daniel Webster, New England's favorite son, rose to defend his section and the American System from attack. He turned the debate from the narrow issue on which it began, however, into a triumphant vindication of the Union against the latest states' rights heresy, nullification as maintained in South Carolina.[9]

8. *Register of Debates*, 21st Cong., 1st Sess., Senate, pp. 118, 22–26, 95–119.
9. *Ibid.*, 33–34 and *passim*.

The issue of the public lands, while narrow, went to the foundations of national policy and was one of the most difficult Congress had to face. In 1821 the Maryland legislature adopted and sent to Congress a report recommending that the public lands be fairly distributed among the states, with the revenue pledged to the purposes of education. Maryland came to the position honestly, for within six months of American independence she had stoutly insisted that the western lands, having been the property of the crown rather than of claimant states, should become the common property of the Union. And with the passage of years she gained her object. The national domain was created. In ceding their western claims, Virginia and the other "landed states" made it a condition that they be held as a common fund for the common benefit of the states. As long as the lands were pledged primarily to the payment of the national debt, there was no quarrel with congressional policy. But Congress repeatedly made large grants in the new western states for the benefit of education and other public purposes. Under the famed Northwest Ordinance, one section of every township was set aside for schools. By 1820 over eight million acres had been appropriated in this way to the so-called public-land states, and it seemed only fair to extend the largess to the original Atlantic states (plus Vermont and Kentucky), which had contributed their people and wealth to the West. Nine states quickly endorsed the Maryland proposition. Although it made little headway in Congress, it raised questions of justice and policy that would not soon be stilled.[10]

The problem became more urgent as the time approached when the national debt would be extinguished at last and the Treasury would actually boast a surplus. Ironically, this blessed state, so long desired, threatened to levy a curse on the country. The debt was a bond of Union. Its retirement would leave a void. And while everyone agreed on the pay-

10. Report Relative to the Appropriation of Public Lands for the Purposes of Education, in *Annals of Congress*, 16th Cong., 2nd Sess., pp. 1772–83. It is discussed in "Appropriation of Public Lands for Schools," *North American Review*, IV (1821), 310–42.

ment of the debt, hardly anyone could agree on what to do with the projected surplus. The public lands netted only a fraction of the annual revenue (something over two million dollars in 1830, about 10 percent of the total), but they were a crucial element in any plan to reduce the surplus because of their special character as a national trust. In that character they had been pledged to the debt; some new commitment, separate from the tax revenue to support the normal operations of government, seemed to be called for.

The policy options with regard to the anticipated surplus were clear by 1830. Not surprisingly they expressed the prevailing interests of the different sections. First, the southern option called for cutting the tariff drastically with a view to reducing the revenue to the level necessary for the economical administration of the government and no more. If the government didn't need the money, if there were no constitutional objects to which it could be appropriated, it seemed only sensible to leave it with the people. Aside from the risks of the policy to domestic manufactures, however, it was less than an adequate solution, since lower duties could not ensure sharply reduced revenue. [11]

Second, the western option recommended giving up the public lands or, at least, disposing of them on such terms as would greatly reduce the income from their sale. Since 1824 Benton had advocated his Graduation Bill under which the price of unsold lands on the market would be annually reduced until, finally, they would be given away. The bill came to a vote and was defeated in 1828. But Benton persevered; some thought he provoked the great debate in 1830 in order to make a bargain for the southern votes necessary to secure passage of the Graduation Bill. Governor Ninian Edwards of Illinois, meanwhile, had boldly put forth the idea of federal cession of the public lands to the states where they lay. Since ownership of the lands was an incident of sovereignty, Edwards argued, public-land states like Illinois were little more

11. See, for example, the remarks of Samuel Smith, in *Register of Debates*, 20th Cong., 2nd Sess., Senate, p. 32.

than tenants of the national overlord. This was an anomaly in
a union founded on the equality of states. Cession, a radical
corrective, appealed to southern states' rights sentiment as
well as to western interests. [12]

The third option—the eastern one—was distribution of the
surplus, or of the proceeds of the public lands (not the lands
themselves, as in the abortive Maryland plan), to all of the
states. The basic idea could be traced back to Jefferson's administration, when the prospect of a debt-free government
first appeared on the horizon, only then to be erased by foreign crisis and war. In his Second Inaugural Address Jefferson
had advocated "a just repartition" of the surplus among the
states for purposes of education and internal improvements. [13]
The Maryland proposition first tied distribution to the public
lands. In 1825 Senator Josiah Johnston, of Louisiana, one of
Clay's political associates, proposed to invest the proceeds of
the public lands in a permanent capital fund, the income
of which would be distributed to the states after each census
to be used for the same purposes. The plan was premature. [14]
In the same year Rufus King, the New York senator and former Federalist chieftain, disclosed some of the potentially
inflammatory effects of the surplus when he advocated chan-
neling the net proceeds into a fund for the compensated emancipation of slaves and their colonization along with free blacks.
Hayne at once countered with a resolution declaring the proposal unconstitutional, dangerous to the slave states, and "calculated to disturb the peace and harmony of the Union." The

12. On the western interest in public lands policy, see Raynor G. Wellington, *The Political and Sectional Influence of the Public Lands, 1828–1842* (N.p.,
1914). An interesting perspective on the cession proposal is William T. Hutch-
inson, "Unite to Divide; Divide to Unite: The Shaping of American Federalism," *Mississippi Valley Historical Review*, XLVI (1959), 3–18.

13. For Jefferson, compare the Second Inaugural and the Sixth Annual
Message, in Paul L. Ford (ed.), *The Writings of Thomas Jefferson* (New York,
1897), VIII, 343, 493–94. See also Albert Gallatin's Report on Roads and Canals
(1808), in *American State Papers, Miscellaneous* (Washington, D.C., 1832–61), I,
724–41, which proposed to use the surplus for internal improvements under
national auspices.

14. *Register of Debates*, 18th Cong., 2nd Sess., Senate, p. 42.

American Colonization Society took up the plan, but it had no future in Congress. [15] More promising was the bill introduced in the Senate by Mahlon Dickerson, of New Jersey, in December, 1826. He proposed a four-year experiment, in anticipation of discharge of the debt, to distribute five million dollars annually among the states for education and internal improvements. Dickerson, a protectionist, assumed that tariff duties should not be generally reduced; but he was also a prototypical Jacksonian Democrat, alarmed by the creation in the Adams administration of a formidable national system of internal improvements "frightful from its immense variety and magnitude," and certain to corrupt republican government. Letting the government collect the surplus, then distribute it to the states seemed the perfect solution. [16] Dickerson's bill was tabled in successive congresses. The same plan, though founded on the public lands proceeds, got a thorough airing in the House of Representatives at the time of the Webster-Hayne debate. [17] Its sponsors were Vermont protectionists. Western congressmen considered it hostile to their states; Benton, in the Senate, denounced the scheme as the "twin brother" of Foot's Resolution. The Charleston *Mercury*, a leading voice of Palmetto State opinion, called it "a system of public robbery devised by Yankee headwork to plunder the South . . . the perfection of the American system." [18] None of the sectional policies could command a majority in Congress.

It was the other oppression, the one traced to the protective tariff and felt most severely in the staple-producing states of the South, that led directly to the Compromise of 1833. The idea of the unconstitutionality of the tariff arose among Virginia states' rights Republicans after 1819. Philip P. Barbour first made the case in Congress in 1824. It was part of a sweep-

15. *Ibid.*, 623. See Robert Ernst, *Rufus King, American Federalist* (Chapel Hill, 1968), 392–93; and Betty L. Fladeland, "Compensated Emancipation: A Rejected Alternative," *Journal of Southern History*, XLII (1976), 170–86.

16. *Register of Debates*, 19th Cong., 2nd Sess., Senate, pp. 209–22. See also the debate in 20th Cong., 2nd Sess., Senate, pp. 25–34 and *passim*.

17. *Ibid.*, 21st Cong., 1st Sess., House of Representatives, pp. 477–504.

18. Quoted in Wellington, *Public Lands*, 25.

ing reaction against postwar nationalism and the progress of "Consolidation," the monstrous movement that in the judgment of these Virginians promised to "reach the whole intercourse among men and even include the connubial."[19] As South Carolina assumed the leadership of the resistance, the tariff became the overwhelming concern. Uniting a theory of free trade with a theory of state sovereignty, and playing on fears of slave revolt and economic decline, the Carolinians began to build an independent party around a single issue unlike anything the country had known before. Such prominent congressmen as George McDuffie and James Hamilton, Jr., repudiated the nationalist opinions they had formerly held; the vice-president, Calhoun, already the weightiest power in South Carolina politics, experienced the same conversion, though he did not confess it publicly for several years. The passage of a still higher tariff, the so-called Tariff of Abominations, in 1828 radicalized the movement. Webster and Clay (who actually had nothing to do with this atrocity) were burned in effigy at Columbia, the state capital; mass meetings of protest were held from Charleston to Abbeville. McDuffie summed up the sentiment in a toast at one of these meetings: "The Stamp Act of 1765 and the Tariff of 1828—kindred acts of despotism!"[20] The tariff, while it was surely not the root cause of southern distress, became the great symbol of oppression because it was so manifestly discriminatory against the South.

In the furor over the new tariff the movement found its rallying cry, its constitutional justification, and its sovereign remedy as well, in the theory of nullification. That fall, while the country was reelecting him vice-president, Calhoun penned an exposition of nullification at the request of a special committee of the state legislature. The whole system of protection, he

19. *Register of Debates*, 18th Cong., 1st Sess., House of Representatives, pp. 916–45. John Taylor of Caroline is quoted. See the discussion in Merrill D. Peterson, *The Jefferson Image in the American Mind* (New York, 1960), 37–39.
20. Washington *National Intelligencer*, August 9, 1828. See also the reports in *Niles' Weekly Register*, June 14 and July 9, 1828. On the South Carolina nullification movement in general, see William W. Freehling, *Prelude to Civil War: The Nullification Controversy in South Carolina* (New York, 1966).

charged, was "unconstitutional, unequal, and oppressive." It was unconstitutional because the congressional power to tax was limited to the object of raising revenue; it could not be used to encourage particular branches of industry or divert the natural course of economic development. (Oddly, Calhoun made no mention of the commerce power, which was part of the protectionists' constitutional defense.) It was unequal because the southern states, by their exports, paid for the bulk of American imports, and therefore of the revenue at the customshouses, yet received virtually nothing in return. "We are the serfs of the system,—out of whose labor is raised, not only the money paid into the Treasury, but the funds out of which are drawn the rich rewards of the manufacturer and his associates in interest. Their encouragement is our discouragement."[21]

Calhoun went on to observe the dangerous tendency of the system, founded on the advancement of sectional interests, to override the restraints of the Constitution and establish a tyranny of the majority. Fortunately, the states, if not the geographical sections, possessed lawful means of defense. The Constitution was a compact of the states, in which each retained its sovereignty and independence. With that went the right and duty of a state to interpose its authority to nullify a federal law deemed in violation of the Constitution. The federal government, being only an agency of the states, could not be the ultimate judge of its own powers. It must concede the disputed power unless three-fourths of the states should grant it by way of amendment. Calhoun, with the nullification party, claimed the high authority of Jefferson and Madison and the Virginia and Kentucky Resolutions of 1798 for this theory. He advocated it as a moderate and peaceful means of preserving the Constitution. The power of veto in each state would be exercised with restraint; indeed, its very existence would introduce comity and conciliation into the affairs of the Union.[22]

21. Crallé (ed.), *Works*, VI, 10.
22. On Calhoun and nullification, see the later chapters in Charles M.

The movement was entirely precautionary in 1828. In December the legislature issued its usual protest against the tariff and published Calhoun's paper, without adoption or endorsement, as the *South Carolina Exposition*. It was hoped, even expected, that Andrew Jackson's election as president would lead to reform of the government, abandonment of protection, and adoption of the revenue standard. He had, after all, swept the southern states as a candidate against consolidation, against the mingled evils of corruption, waste, and privilege charged to the American System and the Adams administration; and the vice-president, although he had been elected as a nationalist four years before, was now identified with the cause of free trade and states' rights.

But the hope of reform was already fading when Hayne, goaded by Webster, expounded "the Carolina doctrine" in the Senate in January, 1830. The new president, for all his posturing about states' rights, had long been a moderate protectionist; as a senator from Tennessee he had voted for the Tariff of 1824; and he had made Pennsylvania, the banner protectionist state, the stronghold of his party in the North. In his first annual message to Congress, Jackson spoke favorably of the existing tariff, recommending only minor corrections. Looking ahead to the extinction of the debt, he thought it unlikely the tariff could be safely reduced to the revenue level, thereby ridding the treasury of a surplus. Nor should the surplus be absorbed in works of internal improvement, since he considered that national policy of doubtful constitutionality. The best solution to these problems, he told Congress, was to distribute the expected surplus among the states.[23] This, of course, was the plan first put forward by Dickerson. It had twice failed to win favor in Congress, and seemed unlikely to do so in the aftermath of the Webster-Hayne debate. Nevertheless, Jack-

---

Wiltse, *John C. Calhoun, Nationalist, 1782–1828* (Indianapolis, 1944), and the earlier chapters in Wiltse's *John C. Calhoun, Nullifier, 1829–1839* (Indianapolis, 1949). See also Augustus O. Spain, *The Political Theory of John C. Calhoun* (New York, 1951), and Peterson, *Jefferson Image*, 51–66.

23. Richardson (ed.), *Messages and Papers*, II, 450–52.

son stuck to distribution and boldly renewed his advocacy when Congress reconvened in December.[24]

Calhoun was deeply disappointed. Distribution, he later said, was the first issue on which he and the president separated, for it announced a design to perpetuate the tariff and complete the ruination of the South. Nothing, not even the tariff, was more odious to the South than distribution, he said. Once the states acquired an interest in the United States Treasury, every one of them, not excepting South Carolina, would become partisans of the tariff. The oppressive system would be fixed permanently on the nation; the states, seduced by plunder, would sink into one universal mass of corruption. The economic exploitation of the protective tariff now appeared as only half the evil. The other half was the political degeneracy that must be the inevitable consequence of distribution, in Calhoun's opinion. This "daring and profligate scheme"—not the tariff alone but the tariff in conjunction with distribution—blasted the hopes of Calhoun and his followers in the Jackson administration and emboldened the nullification movement in South Carolina.[25]

A bitter personal feud completed the rupture between Jackson and Calhoun in 1831. The cabinet broke up in April as a result. When the smoke cleared, it was apparent to everyone that the presidential succession, which had excited so much speculation, was settled not on the vice-president, the early favorite, but on Martin Van Buren, the political wizard from New York who had been Jackson's secretary of state. Southern Jacksonians recoiled from the blow. Angry and disillusioned, some went over to the opposition, carrying their

24. *Ibid.*, 508–17.
25. See Calhoun's letter to C. Van Deventer, May 25, 1831, in J. F. Jameson (ed.), *Correspondence of John C. Calhoun*, Vol. II of American Historical Association *Annual Report, 1899* (Washington, D.C., 1900), 293. Later expressions of disappointment are in his speeches, in Crallé (ed.), *Works*, III, 55–58, 577–78; and in *Life of John C. Calhoun* (New York, 1843), 25–38. Jackson acknowledged his disagreement with Calhoun in writing to John Overton, December 31, 1829, in John S. Bassett and J. F. Jameson (eds.), *Correspondence of Andrew Jackson* (Washington, D.C., 1926–35), IV, 109.

states' rights principles with them. In Virginia the governor
and both United States senators defected; and they began
quietly to advance Calhoun's candidacy for president. Duff
Green, editor of the *United States Telegraph* in Washington,
was the tireless champion of this presidential strategy. So
long as Jackson was not an announced candidate for reelection, the strategy was creditable. Calhoun himself took the
prospect seriously for a time. He had his own "plan of reconciliation" of the great sectional interests. The West would get
its internal improvements, with the sanction of a constitutional amendment; and he would strike a compromise with
the East on the principles of the moderate Tariff of 1816. If
that failed, disunion was inevitable, he said.[26]

No such plan could succeed without the cooperation of
Henry Clay. He was the leading candidate in the field against
Jackson in 1831. His chances, as he recognized, depended on
pulling together the jarring fragments of the opposition, com-
bining with his National Republican following the unruly
Antimasons of the North and the alienated states' rightists of
the South. But there were limits beyond which Clay would
not go, even to defeat Andrew Jackson, and one of them was
cooperation with Calhoun. The veteran statesmen, rivals in
ambition for twenty years, did not like each other; more to
the point, their principles were "directly opposed," as Clay
said, and he dared not risk being tarnished by nullification. Neither Clay nor his foremost eastern advocate, Webster, thought much of Calhoun's influence outside South Carolina. Overtures to Clay from that direction were indignantly
spurned. Although the Jackson press sent up the cry of "coalition," nothing remotely resembling that odious thing mate-
rialized.[27] In the end, Green and a few like him, who would

26. James H. Hammond, "Memorandum," March 18, 1831, in Ulrich B.
Phillips (ed.), "Letters on the Nullification Movement," *American Historical
Review*, VI (1902), 741–45. See also Frederick W. Moore (ed.), "Calhoun As
Seen by His Political Friends . . . ," *Publications of the Southern History Associa-
tion*, VII (1903), 159–69, 269–91.
27. Henry Clay to Thomas Speed, May 1, 1831, in Typescripts, Papers of
Henry Clay, University of Kentucky. See also Clay to Francis Brooke, April 1,

support the devil himself against Jackson, rallied to Clay's standard. Calhoun, on the other hand, acquiesced in Jackson's reelection rather than aid the architect of the American System.

Calhoun's South Carolina friends had considered his presidential prospects hopeless from the beginning. In their view the only viable strategy, after the failure of the Jackson administration to rescue the South, was nullification. Hamilton, now the governor, provided expert political generalship and, with the touch of an impresario, staged balls, parades, and barbeques to arouse enthusiasm for nullification. In May he brought McDuffie to Charleston for a patented oration of fire and fury. The congressman expounded his favorite "forty-bale theory," which held that the protective tariff was the equivalent of a 40 percent export tax on the cotton planters (an argument economically unsound but politically effective), denounced the Union as a foul monster, and called on the brave citizens of South Carolina to throw off their chains.[28] Astonished by this, Green hurriedly inquired if Hamilton, McDuffie, and company "were all crazy." They were ruining Calhoun as a presidential candidate. But that was precisely the Nullifiers' purpose. They wished to attach Calhoun unequivocally to the cause of the state and of the South. His ruin as a presidential candidate was nothing, Hamilton said, compared to the ruin that must ensue from the surrender of principles in a Janus-faced compact between cotton planters and manufacturers, as practical as a confederation of Poles and Cossacks.[29]

Between these feuding friends Calhoun kept his own counsel at his up-country plantation, Fort Hill. McDuffie's tirade took him by surprise. He was still trying to restrain the radicals. Although a Nullifier in theory, he had not pledged himself to nullification in practice—only his most violent Jacksonian enemies accused him of that. He would have liked more

1832, *ibid*. Rumors of coalition are discussed in the *United States Telegraph*, June 3 and 23, 1831, and the Richmond *Enquirer*, March 29, 1831.

28. Speech, May 19, 1831, cited in Freehling, *Prelude*, 221–23.

29. Hamilton to James H. Hammond, June 11, 1831, in Phillips (ed.), "Letters on Nullification," *AHR*, VI, 646–47.

time before choosing between opposing strategies. But Mc-
Duffie's challenge to the pride and patriotism of South Carolina forced his hand. "I see clearly that it brings matters to a
crisis," he wrote, "and that I must meet it promptly and man-
fully."[30] Soon the word went out that he was the author of the
Exposition of 1828. In July he issued the Fort Hill Address,
which committed him to nullification. The editors of the Washington *National Intelligencer*, recalling Calhoun's distinguished
career as a national statesman, expressed their "extreme surprise and deep mortification"; and Green said the address
"was like a shock produced by the cold bath."[31] Calhoun had
burned his bridges to the presidency, not just in 1832, when it
did not matter, but probably forever.

   The polarizing conflict between the administration and the
ascendant nullification party in South Carolina, between Jackson and Calhoun, cried out for a mediator. Clay was the obvious choice. No one had a deeper political interest in the
resolution of the conflict, an interest not only in saving the
American System but in gaining favor in the South. Moreover, Clay had consummate skills for playing the honest broker, as he had demonstrated during his years as speaker of
the House of Representatives and in the negotiation of the
Second Missouri Compromise. He was bold, resourceful, and
persevering; the juggling of interests was his stock in trade.
For the past two years in political retirement at his Lexington home, Ashland, he had been carefully positioning himself to run for the presidency in 1832. Except for the Jackson-
Calhoun rupture, however, things had been going badly for
Clay. The sudden rise of the Antimasonic party in the middle states spread havoc among the National Republicans. Ex-

   30. Calhoun to Samuel Ingham, June 16, 1831, in Jameson (ed.), *Correspondence*, 294. For conflicting opinions on whether Calhoun was a Nullifier, see
the Washington *Globe* editorial, in *National Intelligencer*, April 12, 1831, and the
denial in the *United States Telegraph*, April 9, 1831.
   31. *National Intelligencer*, August 17, 1831; Duff Green to Richard K. Crallé,
August 21, 1831, in Moore (ed.), "Calhoun As Seen," 167. The *National Intelligencer*, June 25, 1831, copying the Charleston *Mercury*, reports that Calhoun
is the reputed author of the *Exposition*.

pected gains from Jackson's embarrassments did not materialize. Even in the South the president remained immensely popular, in part because of such policies as Indian removal and opposition to recharter of the Bank of the United States. With one flourish of the pen, vetoing the Maysville Road Bill and thus striking at the ganglion of federal internal improvements, he surely won more southern votes than he lost by the advocacy of tariff and distribution. The crippling blow to the challenger fell in the August election in Kentucky, when the lowly Jacksonians rose up to smite Clay on his home ground.[32] He was discouraged and mortified as never before. Many of his warmest friends abandoned hope for his candidacy. One of them, Webster, urged Clay to put the presidential contest aside and come into the Senate in the new Congress. "Everything valuable in the Government is to be fought for, and we need your arm in the fight," wrote the Massachusetts senator.[33] Under these persuasions, Clay's spirits revived. He offered himself for the Senate, and the legislature, still at his bidding, elected him in November.

Political tongues at once began to speculate on the course Clay would pursue in Congress. For the past several weeks, since the October meeting of the great Free Trade Convention in Philadelphia, there had been rumors of letters to particular friends in which Clay vowed to seek political accommodation on a moderate tariff. The editors of several National Republican newspapers were soon predicting that he would undertake to lower the revenue of the tariff by the amount of the projected annual surplus (about ten million dollars) but would maintain protection by taking down the duties only from articles not in competition with domestic productions. A long article in the *American Quarterly Review*, supposed to have been written by Clay's senatorial friend, Josiah Johnston, recommended compromise on the same terms. The Richmond *Whig*,

32. Although the National Republicans preserved their legislative majority in 1831, the Jackson party made impressive gains and elected eight of twelve congressmen. In the press the election was widely interpreted as a defeat for Clay and the National Republicans.
33. Webster to Clay, October 5, 1831, in Colton (ed.), *Works*, IV, 318.

thought to be a reliable spokesman, went further, saying that Clay now wished to reduce the tariff to the actual wants of the Treasury, with the duties so arranged as to give only *incidental* protection to the home product. Had Virginia, or Calhoun himself, asked for more? "It is thus manifest," the *Whig* stated, "that the progress of events, and the changed condition of the exchequer have produced a striking coincidence of thought and view between men of supposed different schools, and a near approach of conformity even between extremes."[34]

The *Whig* was generally accurate as to Clay's new position, and it correctly traced the change to the surplus. Renewed predictions of a coalition between Clay and Calhoun were premature, however. The Kentuckian's plan to accomplish the reduction of revenue by lifting duties on noncompetitive articles, many of them luxuries, was at once denounced as a fraud by free traders. He had made no concession of the principle of protection; and while it was true that Calhoun had in a manner conceded *incidental* protection in the *Exposition*— that is, incident to raising revenue—this was not at all the same as Clay's shifty use of that expedient to save the system of protection itself. There was also the matter of Clay's continued adherence to internal improvements under some form of federal financing. This left him all the more vulnerable in the South. Thomas Ritchie, editor of the proadministration Richmond *Enquirer* (one of the most influential newspapers of the time), finally concluded that the conception of Clay as "The Mediator of the South" was a delusion.[35]

When the Twenty-second Congress convened in December, 1831, President Jackson acted at once to check the threatened union of Clay and Calhoun and keep the game in his own hands. He proudly announced that the debt would be

34. Richmond *Whig*, November 23, 1831. See also the *Enquirer*, October 14, November 22 and 27, 1831, and the Alexandria *Gazette* editorial rpr. in the *National Intelligencer*, November 9, 1831. The article attributed to Johnston is "Free Trade and the Tariff," *American Quarterly Review*, X (December, 1831), 444–74. For Clay's statement of position, see his letter to Francis Brooke, in Colton (ed.), *Works*, IV, 314–17.

35. *Enquirer*, November 22, 29, December 10, 1831.

extinguished at the completion of his first term. It therefore became possible, indeed essential, to reduce the revenue to the costs of an economical administration of the government. For the first time Jackson made tariff reform, looking to a revenue standard, administration policy. Cheered by this, Nullifiers and other enemies of the American System were equally pleased by the silent omission from Jackson's message of the plan to distribute the surplus. The president thus smartly anticipated Clay's initiative and at the same time proposed to "annihilate the nullifiers" by removing the pretext of complaint.[36]

From an ideological standpoint the message signaled full-scale retreat from the Madisonian platform toward the denationalization and devolution of public policy that came to characterize the Jackson and Van Buren administrations. Thus far it had shown itself most clearly in Indian affairs. The protecting arm of the federal government had been withdrawn, and removal was forced upon eastern tribes, like the Cherokee in Georgia, at the pain of subjection to state authorities. The rejection of comprehensive federal aid and planning of internal improvements had also manifested the change. Before the end of the present session of Congress, the change became a revolution in the opinion of opposition leaders. It would be dramatized by Jackson's veto of the bill to recharter the Bank of the United States; it would show itself further in land policy and in lesser matters, until it seemed that Jackson was destined to realize his ideal of government as "a simple machine."

In the House of Representatives, consideration of the tariff was divided between two committees, each of which reported its own bill. In the nature of its responsibility the Ways and Means Committee, under McDuffie's chairmanship, was concerned with revenue, not protection. Reporting in February, McDuffie proposed to slash the duties on the principal protected articles to 12½ percent *ad valorem* in two years and ad-

36. Richardson (ed.), *Messages and Papers*, II, 544–58; Jackson to Van Buren, December 17, 1831, in Bassett and Jameson (eds.), *Correspondence*, IV, 383.

mit everything else duty free. The bill embodied the terms of the South Carolina ultraists, yet McDuffie, an austere and angry man, drove it through the committee. The Committee on Manufactures, protectionist in outlook, finally reported its bill in May. The chairman, surprisingly, was John Quincy Adams, the former president, just now beginning a remarkable congressional career in the opposition ranks. He had neither taste nor special qualifications for the committee chairmanship, and seems to have been appointed by the Speaker, Andrew Stevenson of Virginia, in order to embarrass the Nullifiers. Adams considered nullification a kind of "organized civil war," and Calhoun "but a pupil of the Hartford Convention."[37] But Adams had never been an active protectionist (he had expressed his nationalism in foreign policy and advocacy of internal improvements) and had no experience with the mysteries of tariff schedules, though of course as a Massachusetts representative he supported domestic manufactures. Since the majority of the committee were devoted Jacksonians, Adams very soon realized that any bill framed under his auspices, indeed any opposition plan, must fail and that only an administration plan could succeed. So he awaited the lead of the secretary of treasury, Louis McLane, himself a moderate protectionist.

Henry Clay had no patience with this passive approach. In meetings he arranged with Adams and other opposition leaders near the end of December, the senator's charm was equaled only by his arrogance. "Mr. Clay's manner, with many courtesies of personal politeness, was exceedingly peremptory and dogmatical," Adams noted in his diary.[38] Although Clay had served him loyally as secretary of state, he had never approved of the smiling Kentuckian's manners or morals, and in 1831

37. Adams to Clay, September 7, 1831, in Colton (ed.), *Works*, IV, 311–14. See also Charles Francis Adams, "John Quincy Adams in the Twenty-Second Congress," Massachusetts Historical Society *Proceedings*, 2nd Ser., XIX (1905), 509–10.
38. Diary (Adams Papers [Microfilm], Massachusetts Historical Society), December 28, 1831.

the old colleagues had fallen out over the issue of Freemason-
ry. Adams, an ardent Antimason, thought the suppression of
the secret order more important than the election of a president. Clay, the National Republican nominee, formerly a Mason in good standing, thought the furor over the institution
a tempest in a teapot and refused to denounce it in order to
secure Antimasonic support. As to nullification, Adams considered the danger clear and present; Clay thought it bluff.
Adams would not hazard the Union or be obliged to put
down nullification at the cannon's mouth for the sake of the
protective tariff. "I tell gentlemen they must relieve the South
or fight them," he reportedly declared.[39] The South hailed him
as a convert. Clay, despite recent conciliatory statements,
spoke determinedly of maintaining the American System.
Jackson and the Nullifiers had seized upon the burgeoning
surplus as an excuse to destroy it. The way to defeat them and
save the system, Clay thought, was to reduce the revenue immediately, regardless of debt or surplus, by lifting duties on
most nonprotected articles and imposing prohibitive duties
on some protected ones. Adams objected that this defied not
only the administration's plan first to extinguish the debt but
also its aim to meet legitimate southern objections to the protective tariff. To this Clay grandly replied, "I do not care who
it defies. To preserve, maintain, and strengthen the American
system I would defy the South, the President, and the devil."[40]
The two men, unable to agree, went their separate ways.
Adams, never generous in assessing the motives of colleagues, believed "Great Hal" placed his presidential ambitions ahead of the national interest. Always the politician, Clay
sought first to satisfy his constituents in the protectionist
states. And to allow Jackson the glory of extinguishing the
debt or of resolving the conflict with South Carolina was un-
thinkable. "It is an electioneering movement, and this was

39. *Enquirer*, January 24, 1832; Charleston *Courier*, January 21, 1832.
40. Charles Francis Adams (ed.), *Memoirs of John Quincy Adams* (Philadelphia, 1874–77), VIII, 446.

the secret of these meetings," Adams wrote, "as well as of the desperate effort to take the whole business of the tariff reduction into his own hands." He was convinced, moreover, that if Clay succeeded with his plan "blood would flow."[41]

On January 9 Clay introduced in the Senate a resolution for the abolition of duties on all noncompetitive articles except wines and silk, which should only be reduced, and instructing the Committee on Finance to report a bill accordingly. He had obtained the prior approval of Webster and several other protectionists, and felt justified in this move by the failure of the administration (explained in part by McLane's illness) to offer a plan of its own.[42] Two days later he advocated the resolution in a speech that, while temperate and plausible, was exasperating to senators seeking substantial reform. Ignoring the southern threat, he said that the advent of a surplus alone made necessary a reduction of the tariff. An ample revenue, sufficient to support internal improvements, was wanted. Moreover, nothing in the Constitution authorized the federal government to levy taxes for the purpose of distributing the revenue to the states, as Jackson had earlier advocated. His mode of accomplishing the reduction was "undebatable ground," Clay argued defiantly. "It exacts no sacrifice of principle from the opponents of the *American System*, it comprehends none on the part of its friends." By his inflated estimate, it would immediately reduce the annual revenue seven million dollars, which was less than half of the currently best estimate of the surplus. The reduction would be even less if Congress passed two changes in the revenue laws proposed by Clay—one to substitute "home valuation" for the often fraudulent valuation in the exporting country, the other to curtail sharply the long credits heretofore allowed on the payment of duties. These measures, though offered in the name of better and fairer enforcement of the customs laws, had long

41. *Ibid.*, 447–48; Edward Everett to A. H. Everett, January 17, 1832, in Papers of Edward Everett [Microfilm], Massachusetts Historical Society.

42. Clay to Webster and Webster to Clay, January 8, 1832, in *Papers of Daniel Webster* [Microfilm], Dartmouth College.

been favored by most protectionists, who considered them as important as the duties themselves.[43]

Hayne responded with a counter-resolution that called for gradual reduction of the tariff to the revenue standard, which he put at 15 percent *ad valorem*. Compared to McDuffie in the House, he was the soul of moderation and seemed, some thought, to be inviting an amicable accommodation with the administration. He was promptly supported by the octogenarian chairman of the Finance Committee, Samuel Smith of Maryland. Long a spokesman for Baltimore commercial interests, including his own, Smith expressed dismay at Clay's relentless course in the face of impending disunion and civil war. The proposal to resolve the crisis at no cost to anybody, without touching protection, was both shallow and wicked.[44]

Six or seven weeks earlier, when the session began, talk of compromise had filled the air and Clay had been cast in the role of mediator. These hopes had now become desperate. The supposed coalition of Clay and Calhoun to defeat Jackson's nomination of Van Buren as United States Minister to Great Britain did nothing to revive them. The idea of "the Janus-faced coalition"—the coalition of Nationals and Nullifiers—was again floated by the administration through its quasi-official mouthpiece, the Washington *Globe*.[45] The Senate's rejection of Van Buren was later said to have made him president and to have "emasculated" Calhoun, whose casting vote as vice-president brought it about. But whatever the larger implications of this event, it had no immediate bearing on the great issues in Congress. The combination of Clay and Calhoun (and Webster too) against Van Buren was a piece of opportunism. Not even the *Globe*, though it cried coalition,

43. Daniel Mallory (ed.), *The Speeches of Henry Clay* (New York, 1843), I, 586–98.

44. *Register of Debates*, 22nd Cong., 1st Sess., Senate, pp. 186–94. On Smith, see John S. Pancake, *Samuel Smith and the Politics of Business, 1752–1839* (University, Ala., 1972).

45. *Globe*, January 28, 30, 31, February 1, 2, 1832. A third member, Daniel Webster, was quickly added to the coalition, *ibid.*, January 31, 1832. For the denial, see *United States Telegraph*, January 27, February 2, 13, March 16, 1832.

supposed Clay and Calhoun could unite on a tariff settlement. Its argument was that, while they were violently opposed on every issue, they were united in wishing to keep up
the agitation. "Mr. Clay and his system is essential to Mr. Calhoun and his system." And vice versa.[46] In short, said the
Globe, an accommodation would be fatal to the political interests and ambitions of both men, so they had combined against
the administration. At this stage the prospect of cooperation
between them in the coming election had ceased to concern
even the *Globe*. It certainly did not concern Calhoun, who
took no interest in the election, or Clay, who would not risk it.
It had, in fact, become the solitary delusion of Duff Green.

Clay made his major speech, "In Defense of the American
System," during three days in the first week of February.[47]
Friends and enemies alike acknowledged it was a great speech,
one of the greatest in a long and illustrious career. As usual
when Clay spoke, the Senate chamber was packed; fashion-
able ladies crowded the gallery and even took seats on the
floor, while members of the other house, adjourned for the
occasion, occupied the lobby in the rear. The star performer
in this theater—for such the chamber had become—was tall
and rangy, homely yet graceful in appearance, with a prepos-
terously large mouth, mischievous eyes, a seductively musi-
cal voice, a countenance animated to every mood. His whole
body was an expression of earnestness. Clay was an orator of
infinite expedients—now vehement, now subdued; now
beseeching, now deprecating; now pathetic, now sarcastic—
who had to be heard to be appreciated and who, while he
lacked Webster's intellectual power and Calhoun's logical acu-
men, excelled both in the strength and comprehensiveness of
his views.

He began by glowingly contrasting the present prosperity
of the country with its calamitous condition in the years after
1818. The protective tariff was the agent of this change, of
course. From 1824 it had been settled policy. Property and for-

46. *Globe*, February 1, April 25, 1832.
47. Mallory (ed.), *Speeches*, II, 5–55.

tune had been staked on it; it could not be disturbed without violence to the public faith and economic ruin. In defense of the system, Clay gathered up all the arguments he had been making for fifteen years. Free trade was a delusion; if adopted, he warned, calling up republican fears half-a-century in the past, "it will lead substantially to the recolonization of these states, under the commercial dominion of Great Britain."[48] He made much of the economic benefits of the system, in giving employment to labor, in maintaining agricultural prices by the creation of a home market, and in reducing the prices of manufactured goods. Paradoxically, the duty levied to encourage the domestic article did not raise the price to the consumer, as logic seemed to dictate, and as free traders charged, but after a little time lowered it in nearly every instance by making possible improvements of skill and technology and by stimulating competition and supply in the domestic manufacture. (One of Clay's Kentucky associates, carrying this argument to its logical conclusion, which was that many domestic manufactures could now stand on their own, urged him to proclaim the triumph of the American System and submit to lower tariffs.)[49] In reply to the South Carolina planters, Clay denied that the tariff was the cause of their distress (the main cause was competition from fresh lands in the Gulf states), that it imposed an unequal tax burden on them (the tax was borne equally by all consumers), that it reduced their exports (without affecting exports it had created a large home market for cotton), and so on. Having thus disposed of the economic grievance, Clay berated South Carolina's leaders for attacking the democratic rule of the majority and for practicing intimidation and blackmail on the Union. "The danger to our union," he admonished, "does not lie on the side of persistence in the American system, but in its abandonment." For in the latter event, the industry of great states in the North would be paralyzed, their prosperity blighted, "and

48. *Ibid.*, 18.
49. John J. Crittenden to Clay, February 23, 1832, in Crittenden Papers, Duke University.

then, indeed, might we tremble for the continuance and safety of this Union!"[50] With this startling declaration Clay informed South Carolina that two could play at its game. Holding to "no compromise" between protectionism and free trade, he also seemed to place the government in the predicament of incurring disunion from either policy.

The Virginian John Tyler, replying to Clay, said that he "wields an influence over the legislation of Congress . . . more powerful and more controlling than any other man, or set of men, in the country, the manufacturers, and they alone, excepted."[51] Agents of manufacturers—"lobby members," Smith called them—converged on the Capitol and encountered little resistance to their views. The administration, although it had called for tariff reform, held back, as if afraid of opening itself to the charge of yielding to nullification. Tyler and his states' rights friends felt helpless. Calhoun, feeling his own helplessness in the vice-president's chair, glumly complained of "an ignominious and criminal silence" by the president.[52]

The adoption of Clay's resolution was a foregone conclusion. Moreover, he got the reference changed from Smith's committee to the Committee on Manufactures, of which he was himself a member and Dickerson the chairman. In the accompanying parliamentary maneuvers, several western senators, thinking to extort a price for support of the tariff, moved to charge the committee also to inquire into the expediency of reducing the price of public lands and of transferring them to the states. The renewed effort to join these two great issues surprised no one; the commitment by western politicians of public land policy to a body entrusted with the welfare of manufactures astounded everyone. It was absurd, Benton waggishly acknowledged, yet no more absurd, he suggested, than the decision to entrust the finances of the United States to the Committee on Manufactures. There was, after all, a westerner

50. Mallory (ed.), *Speeches*, II, 48.
51. *Register of Debates*, 22nd Cong., 1st Sess., Senate, pp. 359–60.
52. Calhoun to Samuel Ingham, January 13, 1832, in W. Edwin Hemphill (ed.), *The Papers of John C. Calhoun* (Columbia, S.C., 1978), IX, 543–44.

on the committee, Henry Clay. Let him say that he will keep up the price of public lands so as to force poor workers into eastern factories maintained by the protective tariff![53]

The committee promptly reported Clay's resolution in the form of a bill that would reduce the revenue an estimated $5.5 million dollars. After then turning back the move to kill the bill, thus proving his strength, Clay deferred further action pending developments in the House, where revenue bills were supposed to originate anyway. Several weeks later he submitted his report with an accompanying bill on the public lands resolution. The report firmly rejected graduation, preemption, cession, and similar proposals as a wasteful misappropriation of a precious resource. The national domain—over one billion acres of land—was a "sacred trust," sealed by the blood of the Revolution. It was not only a bond of union but, under the present land system, a guarantor of freedom and progress for countless generations to come. He would have preferred to make no change in the system, but the problem of the surplus suggested the wisdom of a five-year experiment in distribution, under which the states would receive the income from the sale of these lands (about three million dollars in 1831) for purposes of education, internal improvements, and colonization of free blacks. The constitutional and other objections he had expressed to the distribution of the surplus revenue from taxes did not apply to the proceeds of public lands that were in the nature of a trust. Clay's plan would award large bonuses to the seven new states, with millions of federal acres and few people; the rest of the income would be shared by all twenty-four states on the basis of federal population.[54]

With this report Clay believed he had turned a flagitious scheme to hurt him, "to place in my hand a many edged instrument which I could not touch without being wounded," into a brilliant triumph.[55] The basic idea of the Land Bill, as it

53. *Register of Debates*, 22nd Cong., 1st Sess., Senate, pp. 631–35, 638.
54. The report is *ibid.*, Appendix, 112–118. See also Clay's speech, in Mallory (ed.), *Speeches*, II, 56–85.
55. *Register of Debates*, 24th Cong., 1st Sess., Senate, pp. 51–52.

came to be called, did not originate with Clay, but he was the first to develop it into a comprehensive and practical plan. It was a remarkably creative response to new imperatives of public policy. In part, of course, Clay's motives were narrowly political: to turn back the challenge, spearheaded by Benton, to his claim on western suffrages. In this respect, said a leading Jacksonian newspaper, "it is doubtless the boldest move that ever was made by a candidate for the Presidency."[56] The senator was equally concerned not to alienate his eastern constituency, however. The direct payoff would be in the form of annual dividends from the trust, in which all states would share. But the eastern states would be the principal beneficiaries of the indirect payoff through the protective tariff, which would be kept up, since the Treasury would be drained of its income from the sale of public lands. "It is a tariff bill; it is an ultra tariff measure," Benton thundered. "Tariff is stamped upon its face, tariff is emblazoned upon its borders; tariff is proclaimed in all its features."[57] The doughty Missourian even complained of the impropriety of the Committee on Manufactures originating such a bill! It was tabled, and the senators referred the subject to the Public Lands Committee, which subsequently reported a bill to reduce the minimum price of public lands from $1.25 to $1.00 an acre and, on the graduation principle, to only fifty cents for lands on the market five years. The issue was thus clearly drawn. Near the end of the session, after the matters of the bank and the tariff had been settled to Clay's satisfaction, the Senate also passed his Land Bill. But the pressure of other business caused its postponement in the House.

Turning back to the House, it was not until the end of May that the tariff debate began. McDuffie's bill was first disposed of. The congressman, himself a great cotton planter, again argued that the tariff amounted to a 40 percent excise on the export of the southern staples, cotton, rice, and tobacco. If continued, the South must rapidly decline and, as the value

56. Frankfort, Ky., *Argus of Western America*, quoted in Wellington, *Public Lands*, 38.
57. *Register of Debates*, 22nd Cong., 1st Sess., Senate, p. 1151.

of slave labor fell, face the disaster of abolition. This terror, and the terror of northern despotism with all the oppressions of the American System, would be lifted from the southern states if they were outside the Union. McDuffie solemnly concluded that unless Congress provided the remedy, South Carolina, leading the way for her sister states, would nullify the tariff within five months of adjournment.[58] Nathan Appleton, of Massachusetts, a great cotton manufacturer, replied to McDuffie in a brilliant maiden speech in which he was coached by Daniel Webster. Appleton argued learnedly that the tariff as perceived by South Carolina simply did not exist; she was preparing to risk disunion and civil war for "a mere figment of the brain."[59] On his motion the House voted two to one to strike the enacting clause from McDuffie's bill.

The bill from Adams' committee now came before the House. Basically it was Treasury secretary McLane's work. Adams had decided early in the session, it may be recalled, to let the administration frame its own bill. To the astonishment of the National Republicans he had rejected Clay's leadership and looked to Jackson's because he trembled for the Union. McLane's illness (a long and violent attack of gout), compounded by the administration's dilemma on the tariff, delayed his report for several months. Received by the Committee on Manufactures near the end of April, the report offered enough concessions to rally the Unionists in South Carolina and hold the loyalty of the other planting states without submitting to the Nullifiers. It also sought by its moderation to reassure northern manufacturing interests. Indeed, the bill to which the report gave rise was written as if Clay had been looking over McLane's shoulder. It made no across-the-board reduction of duties. It retained the odious minimum on cheap cotton textiles, and most protected articles suffered only slight reductions, if any. As in Clay's plan the principal reductions —generously estimated at ten million dollars altogether— were on nonprotected articles: tea, coffee, tropical produce

58. *Ibid.*, House of Representatives, pp. 3119–70.

59. *Ibid.*, 3190–3209; Appleton to George Ticknor, February 9, 1853, in Nathan Appleton Papers, Massachusetts Historical Society.

generally, tin, flax, dyes, and so on. The main change occurred on wool and woolens, the most perplexing problem of the system for the past several years. Out of it had come the Tariff of Abominations, in 1828, which this bill would repeal, and a burgeoning woolens industry in the eastern states. Now specific duties on imported woolens and the schedule of minimums (wherein the duty was levied at a minimum value regardless of lower prices on the invoice) were abolished. In partial compensation to the manufacturers, coarse raw wool was admitted duty free. And the duty on coarse woolens, which southern planters were believed to use to clothe their slaves, was lowered to 5 percent of value. Other woolens carried duties ranging from 10 to 50 percent. In Adams' view these reforms met the most reasonable objections of the South to the tariff and ought to be accepted as liberal concessions. Taken as a whole the bill he submitted to the House returned the tariff to the level of 1824.[60]

Moderate though the bill was, it immediately created alarm in the industrial Northeast. In the cities "friends of the American System" held meetings of protest. At one of these meetings in Massachusetts the protesters, as if to back up Clay's warning, declared secession from the Union "preferable to the sacrifice of the principle of the protective system."[61] The bill was seen as ruinous to woolens manufactures, in which large investments had only recently been made. Confidence in the permanency of protection was so badly shaken that mills were already closing, it was reported. The Rhode Island legislature, then in session, instructed the state's senators, and requested the representatives, to oppose the McLane-Adams

60. McLane's report and also the report of the Committee on Manufactures are in *Register of Debates*, 22nd Cong., 1st Sess., House of Representatives, pp. 25–33, 79–93. For South Carolina congressman William Drayton's view of the bill, and comments on McLane's illness, see his letters of April 5 and May 2, 1832, to Joel R. Poinsett, in Poinsett Papers, Historical Society of Pennsylvania. See also John Munroe, *Louis McLane: Federalist and Jacksonian* (New Brunswick, 1973), 310–14.

61. *Boston Courier*, June 15, 1832.

bill.[62] Rufus Choate, a Massachusetts congressman, summed up the reaction: "Not one *manufacturer* believes we can live under it." Yet he feared the bill would pass. "The delusive cry of 'Union in danger,' 'compromise,' etc. will rally timid and confused minds who can't see that half protection—inadequate protection—is as useless as half a pair of scissors—or as to be almost a Christian."[63] The Nullifiers thought no better of the proposal. It was, said the Charleston *Mercury*, a scheme to save northern manufacturers, offering no relief to the South. Regardless of the bill's fate in Congress, the South Carolinans were resolved to take the tariff to the test of nullification.[64]

The House passed the bill essentially unchanged on June 28. The vote was surprisingly lopsided—132 to 65. It was a victory of the broad middle against the extremes. The nays were almost evenly divided between North and South, protectionists and free traders, Nationals and Nullifiers. The Massachusetts delegation voted 8 to 4 *against* passage, the South Carolina delegation 6 to 3. Those three votes for the bill were exceedingly important as an expression of Unionist feeling in the state. The vote of the southern states as a whole favored the bill, which revealed the hollowness of the South Carolina claim to leadership of the section and buoyed the administration's confidence that the state would stand alone in nullification. Many of these southern votes (all nine of Tennessee's), it seems safe to say, were halloos for Jackson. As a New England congressman remarked, "It gives the death blow to Nullification, but I fear strengthens Jackson."[65] In northern state delegations strongly under Jacksonian influence, the bill assumed the character of a party measure. This was true of New Hampshire, for instance, and alone can account for the unanimous vote of its representatives for the bill. It was probably true

62. *Enquirer*, May 22, 1832. See also Jeremiah Mason to Webster, May 22, 1832, in Webster Papers.
63. Rufus Choate to Jonathan Shove, June 3, 1832, in "Rufus Choate Letters," *Essex Institute Historical Collections*, LXIX (1933), 81–82.
64. *Enquirer*, April 17, May 13, 1832.
65. Edward Everett to A. H. Everett, July 1, 1832, in Everett Papers.

of Pennsylvania, a state both protectionist and Democratic, which split its vote. But New York offered the more convincing case. It favored the McLane-Adams reform 30 to 2. Despite the substantial body of free-trade opinion in the state, centered in New York City, almost all Democratic, and despite its ties to Virginia ideologues, the party was predominantly protectionist. Van Buren was still in England, so the delegation's nearly unanimous vote for the bill cannot be directly credited to his fine political hand. His associates in the Albany Regency had surely not been idle, however, and his vice-presidential candidacy doubtlessly inspired loyalty to the administration.

In the Senate, the Committee on Manufactures, to whom the House measure was referred, reported the bill with a long list of amendments, nearly all of them protectionist. Day after day in the torrid July heat of the capital, at the tired end of one of the longest sessions on record, the Senate debated and approved most of the amendments. The protectionists' effort to restore the minimum on woolens was defeated, but they succeeded in raising the duty at the top from 50 to 57 percent. Clay and Kentucky did not get all the protection they wanted for cotton bagging (a hemp product), and this was considered a concession to the South. In the end, however, Clay was satisfied with the amended bill. It had become *his* bill, more than McLane's or Adams', settled basically on the principles he had announced six months earlier.[66] Hayne concurred, sadly, in this judgment. A measure supposedly conceived in the spirit of conciliation, and under the necessity of reducing the surplus, had become the consummation of the American System. "It is neither more nor less than the resolution of the gentleman from Kentucky reduced to the form of law," Hayne said.[67]

On July 9 the Senate approved the amended bill 32 to 16.

66. Clay to Hezekiah Niles, July 8, 1832, in Typescripts, Clay Papers. And see his remarks on July 4 in Washington, in *National Intelligencer*, July 7, 1832.
    67. *Register of Debates*, 22nd Cong., 1st Sess., Senate, p. 1217. See also the

The proportion duplicated the House vote, but the division in the Senate was sharply sectional. Every northern senator but one (Kane of Illinois) voted yea. Every southern senator voted nay, except Clay, the senators from the border states of Maryland and Missouri, and those from the sugar-producing and tariff-dependent state of Louisiana. The amendments were entrusted to three senators appointed to confer with their counterparts from the House. To the disgust of Clay, Webster, and several others, the chairman of the Senate conferees, William Wilkins of Pennsylvania, brought back a recommendation to recede from the amendments. Wilkins was a protectionist, especially where iron was concerned, but he was also a Jacksonian Democrat who harbored vice-presidential ambitions; and it was to this that his shameless appeal to the South and betrayal of the Senate was attributed. Exhausted, the senators voted to recede amendment by amendment; nevertheless, every protectionist, including Clay and Webster, voted for final passage of the bill.

In some quarters the Tariff of 1832 was hailed as a compromise, perhaps even a permanent settlement of the vexatious issue that endangered the Union. This was the hopeful view of the South Carolina Unionists and of many northern Jacksonians. But the act was not a compromise. One of the principal parties in the contest, the Nullifiers of South Carolina, gained nothing of value and denounced the act as monstrous. Neither of the other principals, the administration as represented in President Jackson, or the American System and the National Republicans as represented in Senator Clay, surrendered anything of value, unless possibly the woolens interest. The president played an aloof, even enigmatic, role throughout the proceedings. In an election year he seemed as little inclined as his rival to risk the loss of protectionist votes. Although cast in the role of mediator, Clay, of course, again proved himself the leading protagonist of the protective tariff. He claimed

---

speech of Georgia Senator John Forsyth, after Congress adjourned, in *Enquirer*, September 11, 1832.

to have saved the American System and, to Adams' chagrin, took most of the credit for the legislation.[68] As the act was not a compromise, neither was it a permanent settlement. It did not meet the problem of the surplus. Reliable estimates of the reduced revenue under the act ran from three to five million dollars. (The new duties, it should also be noted, were prospective, to go into effect March 4, 1833, after the scheduled retirement of the debt, as Jackson had recommended.) It did not materially reduce protection. McDuffie contended that the act, by throwing nearly all the duties on protected articles, actually increased the toll on those articles by $1.5 million dollars. However this may be, it is clear that the legislation failed to deal responsibly and effectively with the critical problems that brought it into being.

The Tariff of 1832 at once became the firebrand of agitation for the South Carolina Nullifiers. After this defeat, they told the people, it was hopeless for the state to look to the federal government for justice. Day after day presiding in the Senate, listening to the tariff debate, Calhoun felt a mounting sense of desperation. The two sides, instead of converging, as they ought to do in the deliberative process, were driven farther and farther apart. "It is, in truth, hard to find a middle position," he observed philosophically, "where the principle of protection is asserted to be essential on one side, and fatal on the other. It involves not the question of concession, but surrender, on one side or the other."[69] He was determined that the surrender would not forever be on his side. Even before the Senate passed the amended tariff bill, Calhoun set out for South Carolina. "The question is no longer one of free

68. On Adams, see his letter to Louisa C. Adams, July 14, 1832, in Adams Papers. Calvin Colton, *Life and Times of Henry Clay* (New York, 1846), II, 218, simply assumes Clay was the author; Carl Schurz, *Life of Henry Clay* (Boston, 1887), I, 365, credits the bill mainly to Clay. Pennsylvania's junior senator thought his state, and especially Wilkins, deserved the credit. See George M. Dallas to Henry D. Gilpin, July 13, 1832, in the Dallas Papers, Historical Society of Pennsylvania.

69. Calhoun to Francis W. Pickens, March 2, 1832, in Clyde N. Wilson (ed.), *The Papers of John C. Calhoun* (Columbia, S.C., 1977), XI, 558.

trade, but of liberty and despotism," he wrote to a compatriot at home. "The hope of the country now rests on our gallant little State. Let every Carolinian do his duty."[70] The message sounded like a communique—and a declaration of war.

70. Calhoun to Waddy Thompson, July 8, 1832, *ibid.*, 604.

II The Compromise

With the passage of the Tariff of 1832 nullification became a political crusade in South Carolina. The declared
aim of the Nullifiers, who sported the blue and orange cock-
ade of the State Rights and Free Trade party, was to annul and
arrest the tariff within the state's borders, thence to force a
constitutional revolution in the Union along the lines of John
C. Calhoun's theory. The South Carolina Unionists, and political observers throughout the country, were skeptical of these
professions, however. Some believed the real object of the agitation was secession and the birth of a southern confederacy.
Most thought that nullification was political bluff, part of an
elaborate strategy of confrontation to extort concessions from
the federal government and perhaps gratify the craven political ambitions of Calhoun and his friends, but nothing else.
The Unionists defended the new tariff as a step in the right
direction, for they, too, were free traders, and supposed it
would yield to further improvement once the protectionist
pressures of the presidential contest between Jackson and Clay
were lifted. The election would also diminish the motive of
Calhoun and his clique to embarrass the Jackson administration.[1] In the Unionist view, the rhetoric *ad terrorem* of the
South Carolina agitation had a conventional political goal, not
the revolutionary one conjured up in Calhoun's theory, and
certainly not the disunionist one of a few Hotspurs.

The complacency of the Unionists stemmed, in part, from

1. On the latter point, see Joel R. Poinsett to Andrew Jackson, November
16, 1832, in Papers of Andrew Jackson [Microfilm], Library of Congress. The
Charleston *Courier* voiced its approval of the new tariff July 23, 1832. The exchange between William Drayton and Robert Hayne on the act is interestingly discussed in the Richmond *Enquirer*, September 11 and 14, 1832. Wil-

their inability to credit nullification as a constitutional theory
or remedy. Professing to save the Union, it would destroy it.
Professing to secure peace and order by restoring the limited
government of the Constitution, it would provoke civil war.
In August Calhoun undertook to refute these objections and,
for the first time, to explain the practical workings of the theory in some detail. Proceeding from the fundamental premise
of state sovereignty, one and indivisible, he argued that it
was the right and duty of a state to interpose its authority
against the unconstitutional acts of the federal government.
One of two courses must follow. First, the government would
acquiesce and abandon the disputed power. That government,
being only an "agent," had no right to coerce one of the "principals" to the contract that established it; even if madness
prevailed and force were attempted, it would be voided by
the courts. Second, a convention of the states would be called,
and if three-fourths of them granted the contested power by
amendment of the Constitution, the nullifying state would acquiesce or, in the last extremity, secede from the Union. But
the aim was to prevent secession. The process of nullification would in fact, Calhoun argued, preserve the Union by
giving ascendancy to the constitution-making authority over
the law-making authority. The latter operated on the rule of
an "absolute majority," which, left to itself, must end in consolidation and tyranny; while the former operated on the rule
of a "concurring majority," in which the states were the active
elements and amendment the means of preserving the Constitution.[2] The theory was ingenious, as Unionists acknowledged, and speciously laid on the groundwork of the Virginia
and Kentucky Resolutions; but it was at war with reason, republicanism, and the Constitution. By what right, they asked,
could one state stop the wheels of government and demand a
convention of all the states? Why, in convention, should one-

---

liam H. Freehling, *Prelude to Civil War: The Nullification Controversy in South
Carolina* (New York, 1966) is the best account of the controversy in South
Carolina.

  2. Calhoun to Governor Hamilton, August 28, 1832, in Richard K. Crallé
(ed.), *Works of John C. Calhoun* (New York, 1853–55), VI, 144–93.

quarter of the states be allowed to decide basic constitutional questions? The result of the whole, as Daniel Webster had observed, was "that, though it requires three-fourths of the states to insert any thing in the Constitution, yet any one can strike any thing out of it." The theory stood exposed, said the Charleston *Courier*, "in all the deformity of anarchy and misrule."[3] Although trumpeted as a peaceful and constitutional remedy, nullification was a fraud on the people of South Carolina.

Whether it was or not would largely depend, if brought to a test, on President Jackson. He was a violent man, of course. No one had better reason to know that than Calhoun. A year before, writing to a Fourth of July Unionist rally in Charleston, Jackson had declared his intention to enforce the law at all hazards.[4] Now he seemed pleased with the new tariff and even hoped it might become the basis of a permanent settlement. His signature on that bill together with his veto of the bank bill had, he said, broken the dangerous coalition of Clay, Webster, and Calhoun; and if the third member of this combination continued to advocate nullification after the concessions of the new tariff, even his own people would see it for what it was, a scheme of "disappointed ambition." The South Carolina movement was incomprehensible to Jackson in any other terms than personal malice and the conspiratorial design of "unprincipled men who would rather rule in hell, than be subordinate in heaven," which practically ensured a violent response.[5] The Nullifiers were somewhat apprehensive, as suggested by a defiant yet disquieting Fourth of July toast— "Andrew Jackson: On the soil of South Carolina he received

3. Charleston *Courier*, August 4, 6, 11, 1832. Webster is quoted from a speech in New York on March 10, 1831, in *The Writings and Speeches of Daniel Webster* (Boston, 1903), II, 60–61.

4. Jackson to John Stoney and others, June 14, 1831, in Jackson Papers.

5. Jackson to John Coffey, July 17, 1832, in John S. Bassett and J. F. Jameson (eds.), *Correspondence of Andrew Jackson* (Washington, 1826–35), IV, 462. On Jackson's outlook, see Richard B. Latner, "The Nullification Crisis and Republican Subversion," *Journal of Southern History*, XLIII (1977), 19–38.

an humble birthplace. May he not find in it a traitor's grave!"[6]

Other signs pointed to a different outcome, however. During four years the president had steadily retreated from the moderately nationalist platform on which he began. In the spring he had backed Georgia's meditated resistance to the decision of the United States Supreme Court in the case of *Worcester* v. *Georgia*. Even if he never uttered the famous line later attributed to him, "Well, John Marshall has made his decision: now let him enforce it!" he said other things in the same spirit.[7] The case concerned Georgia's sovereign claim, which the court rejected, to jurisdiction over the Cherokee territory within the state's limits. Because the claim lay at the foundation of Jackson's policy of Indian removal, his support of Georgia was hardly surprising and did not automatically implicate him in a radical states' rights position. Still it gave encouragement to the Nullifiers and embarrassment to their opponents. If Jackson acquiesced in Georgia, on what basis could he uphold federal authority in South Carolina? Some Unionists were alarmed. "The *old man*," one of them observed, "seems to be more than half a Nullifier himself."[8]

This might have been the theme of Webster's address to the National Republican Convention in Worcester, Massachusetts, in October. Since his celebrated triumph over Hayne, Webster had assumed the public role of Defender of the Constitution. Now, on this partisan platform, he declared the Constitution in "imminent peril" from President Jackson. After noticing the Georgia case and the movement in South Carolina, he argued that the administration had denounced and discarded most of the leading powers of the Constitution that had matured during forty years. And so the national bank had been

6. *Enquirer*, July 20, 1832.

7. See the important article by Edwin A. Miles, "After John Marshall's Decision: *Worcester v. Georgia* and the Nullification Crisis," *Journal of Southern History*, XXXIX (1973), 519–44.

8. James L. Petigru to Hugh S. Legaré, October 29, 1832, in James P. Carson (ed.), *Life, Letters and Speeches of James Louis Petigru* (Washington, D.C., 1920), 102–105.

put on the road to extinction, national planning and support
of internal improvements had been virtually terminated, dis-
mantlement of the public-land system had begun, even the
protective tariff faced an uncertain future. In vetoing the recharter bill Jackson had denounced not only the Bank of the
United States but all national legislation that, in his words,
"arrayed section against section, interest against interest, and
man against man, in a fearful commotion which threatens to
shake the foundations of our Union." Here, in false fears giving rise to false principles, said Webster, was the whole creed.
In his own creed, of course, the Union would be preserved
not by abandoning the national system of legislation but by
strengthening it. Jackson had also questioned the ultimate authority of the Supreme Court to interpret the Constitution,
saying that every officer of the government was bound by oath
to uphold the Constitution "as he understands it." Webster
stood amazed at the assertion of this "wild and disorganiz-
ing" doctrine in a critical time. "Are we not threatened with
dissolution of the Union?" he pleaded. "Are we not told that
the laws of the government shall be openly and directly re-
sisted? . . . Mr. President, I have very little regard for the law,
or the logic, of nullification. But there is not an individual
in its ranks, capable of putting two ideas together, who, if
you will grant him the principles of the veto message, can not
defend all that nullification has ever threatened." There was
unfortunately, he concluded, no evidence that Jackson opposed nullification. And if he should oppose it in the event,
he lacked the principles to oppose it successfully. [9]

In the fall South Carolina advanced in measured strides toward nullification. The legislative election was, in effect, a ref-
erendum on the question. The Nullifiers won an impressive
victory over the Unionists. Governor Hamilton immediately
summoned this newly elected legislature. All hope of a returning sense of justice in the general government had "finally and forever vanished," Hamilton announced; therefore,
he called for the election of a convention, the embodiment of

9. *Writings and Speeches of Daniel Webster*, II, 87–128.

the state's sovereignty, to nullify the tariff. The convention bill quickly passed. After a quiet two-week campaign, in which the Unionists conceded the outcome, the delegates gathered at Columbia. On November 24 they enacted an ordinance nullifying the tariff acts of 1828 and 1832. Its effective date, when the collection of duties would be prohibited, was put off until February 1, however. Resistance, meanwhile, would be legal and voluntary. Not only was this a victory of moderates over radicals but it declared, at this late hour, that South Carolina had not slammed the door on a political solution at the national level. What were the terms of such a solution? Basically, as stated in one of the convention addresses, they were the same as McDuffie had proposed in Congress: an *ad valorem* tariff for revenue only at an average rate of about 12 percent. This was the negotiating position. The ordinance further decreed that any effort by the federal government to coerce the state in order to enforce the tariff in South Carolina would be just cause for secession. Within a few days the legislature reconvened to pass the laws necessary to implement the ordinance—to raise an army, purchase arms, impose an oath of allegiance to the state, and so on.[10] The legislators also elected a new governor, Hayne, and to succeed him a United States senator, Calhoun, who resigned the vice-presidency.

In this enveloping crisis Congress came into session. The president's annual message was eagerly awaited, not only because of events in South Carolina. Jackson had been reelected with ease, winning seventeen states to Clay's six. (South Carolina threw away its electoral vote.) By interpreting the result as a mandate, he could impose his own stamp and style on the public policy of the country with little concern for the effect on his political fate. In an apparent nod to the South, the message called for further reduction of the tariff to the revenue standard. The advantages of protectionism, as in the encouragement of domestic manufactures, were offset by the attendant evils of jealousy, discontent, and disunionism. He

10. The important documents are collected in *State Papers on Nullification* (Boston, 1834).

thought its benefits should be limited ultimately to articles of military necessity. The report of the secretary of treasury, echoing these views, recommended lopping off an additional six million dollars of revenue, which was the estimated amount of the annual surplus that would be produced by the 1832 act. While nullification was a factor in this recommendation, it was not the only one. Setting forth the goal of a government reduced to "that simple machine which the Constitution created," Jackson embraced the tariff in a general plan of denationalization, including government divestiture of stock in the Second Bank of the United States, abandonment of internal improvements, and a new policy for the rapid disposal of public lands ending in their surrender to the states.[11]

These negative clauses of the Jacksonian creed, more than the remarkable complacency of the message before the challenge of South Carolina, disturbed the National Republicans. John Quincy Adams thought it the most deadly blow ever struck to the Union. "It goes to dissolve the Union into its original elements, and is in substance a complete surrender to the nullifiers of South Carolina."[12] Could Webster have been right? The editors of the *National Intelligencer* clearly thought so. Jackson proposed to give up every national position in the face of nullification, they said. One of the Washington correspondents summed up the gloomy response of the opposition press to the message: " 'All—all is gone if the President's views are carried into effect. The Bank is gone! The American System is gone! Internal Improvements are gone! The public lands are gone! All is gone which the General Government was instituted to create and preserve.' "[13]

11. J. D. Richardson (ed.), *A Compilation of the Messages and Papers of the Presidents, 1789–1897* (Washington, D.C., 1907), II, 591–606; Report of the Secretary of Treasury, December 5, 1832, in *Register of Debates*, 22nd Cong., 2nd Sess., pp. 33–39.

12. Diary (Adams Papers [Microfilm] Massachusetts Historical Society), December 5, 1832.

13. Quoted in *Enquirer*, December 11, 1832; *National Intelligencer*, December 8, 1832.

The appearance, only six days later, of Jackson's Proclamation to the People of South Carolina surprised nearly everyone. It was boldly nationalistic—a complete turnabout from the message. So great was the disparity that some of Jackson's opponents suspected him of deceitfully waving the charms of the proclamation before northern eyes while preparing to concede all the demands of South Carolina.[14] Jackson never explained the disparity, or acknowledged its existence. In theory, of course, there was no contradiction between belief in the supremacy and indivisibility of the Union and belief in a federal government of severely limited powers; indeed, Jackson had become convinced that one was necessary to the other. That the Union would be destroyed by consolidation and preserved by adherence to states' rights and strict construction of the Constitution was an old republican philosophy, apparently as congenial to Jackson as to Calhoun. Practically, he had wanted to make a show of conciliation and forbearance so as to hold the loyalty of South Carolina's sister states; and when none rushed to her aid he knew that the South— the whole country—was ready for the rigors of his proclamation. Scornfully rejecting the claims of South Carolina to nullify federal laws, Jackson espoused a theory of the Constitution and Union that was associated with nationalists like Webster and Marshall. The people of the United States, although they had acted through state conventions, created a collective union, a government of *one* people," to which they committed certain sovereign powers heretofore belonging to the state governments; and no state could violate this union or secede from it without dissolving the whole. It was, in short, a binding compact, not a terminable one, which was the fundamental difference between the government of a league and the government of a nation. Concluding with a paternalistic plea to the people of his native state, Jackson sought to separate them from their leaders who, bent on disunion, were rushing the state to certain ruin. "Disunion by armed

14. Boston *Courier*, December 22, 1832.

force is treason," he warned. Force would meet force. The laws of the United States would be executed. The Union would be preserved.[15]

It was said that Jackson's personal appeal, which he likened to "that a father would use over his children," produced a wave of sardonic laughter in the South Carolina legislature when the proclamation was read.[16] The resolutions that were promptly adopted labeled this show of affection, after the undisguised hostility of the proclamation, revolting. The resolutions declared the proclamation itself unconstitutional, charged that its doctrines must produce executive tyranny on top of majority tyranny, reasserted the primacy of state allegiance together with the right of peaceable secession, and concluded defiantly "that the state will repel force by force and, relying upon the blessings of God, will maintain its liberty at all hazards."[17]

The legislature's actions were more cautious than its words, however. Under mounting Unionist pressure, it softened the oath of allegiance mandated by the convention; and while thus fending off the danger of civil war within its own borders, the state was also anxious not to appear the aggressor in any armed conflict with the Union. Even as it raised a volunteer army, the government carefully kept nullification in the legal avenues of writs, judges, and juries. Eventually, of course, when decisions of state courts upholding the nonpayment of duties on imports were appealed to federal courts, there must be a clash; but it was not imminent in December.

Jackson was just as anxious not to strike the first blow. The doctrines and the language of the proclamation caused uneasiness among his closest advisers, including Martin Van Buren, the vice-president-elect, who had labored for years to restore the Jeffersonian "alliance of southern planters and plain re-

15. Richardson (ed.), *Messages and Papers*, II, 1203–19.

16. James W. Wylie to Messrs. Polk and Standifer, January 11, 1833, in Paul H. Bergeron (ed.), "A Tennessean Blasts Calhoun and Nullification," *Tennessee Historical Quarterly*, XXVI (1967), 385–86.

17. Henry S. Commager (ed.), *Documents of American History*, (3rd ed.; New York, 1946), I, 268–69.

publicans of the North." This mandated a states' rights platform for the Democratic party. But every "national" from
Maine to Louisiana approved of Jackson's proclamation. Webster, the most prominent, hailed the document as a vindication of his own preachings. At a Boston "union meeting" he
pledged his unqualified support to the president. The prospect of an old Federalist like Webster becoming Jackson's political bedfellow under cover of nullification was exceedingly
mortifying to Van Buren and his friends.[18]

The administration moved quickly to resolve the crisis
through further tariff reform. The olive branch, it was hoped,
would obviate the sword. Secretary of Treasury McLane col-
laborated with the newly appointed chairman of the House
Ways and Means Committee, Gulian C. Verplanck, to produce an acceptable bill. The complexion of the committee had
not changed from the previous session; and Verplanck succeeded naturally to the chairmanship when McDuffie turned
up absent. A fourth-term congressman from New York, one
of Van Buren's bucktails, and a litterateur to boot, Verplanck
was a committed free trader, though oddly enough he was
probably best known for his defense of the constitutionality
of the protective tariff against the South Carolinians. The bill
he reported at the end of December, with the support of all
but one member of the committee, would slice another six or
seven million dollars from the revenue, as McLane had proposed, and return the tariff to the general level of 1816 when
the present system began. (That first protective tariff was, curi-
ously, still remembered as "the South Carolina tariff" because
of the patronage it had received from the state's leaders.)

The administration immediately took up the Verplanck bill
as its own. It was everywhere considered more Van Buren's

18. *The Autobiography of Martin Van Buren*, Vol. II of American Historical
Association *Annual Report, 1918* (Washington, D.C., 1920), Chap. 36. See also
Amos Kendall to Van Buren, November 10, 1832, and C. C. Cambreling to
Van Buren, December 18, 1832, in Papers of Martin Van Buren [Microfilm],
Library of Congress. Webster's speech, December 17, 1832, is in *Writings and
Speeches of Daniel Webster*, XIII, 40–43. See also Sydney Nathans, *Daniel Web-
ster and Jacksonian Democracy* (Baltimore, 1973), Chap. 2.

than Jackson's bid to settle the crisis, however. Although the terms of the bill fell short of South Carolina's demands and did not actually repudiate protectionism in principle, the proposition might have been acceptable to the Nullifiers but for its auspices. The importance of this issue was magnified when the state chose Calhoun to manage its cause in Washington. The personal enmity between him and Jackson, together with his unreasonable grudge against Van Buren, would make it exceedingly difficult for Calhoun to grasp any olive branch held out by them. Of course, the friends of the American System throughout the northern states opposed the Verplanck bill from the start. They were stunned by the broad sweep of the measure as well as by the swiftness with which it would be accomplished. Protection would be withdrawn in just two years, subjecting many businesses to almost certain ruin. The rejection of gradualism, a feature of earlier southern plans to end the system, reflected the administration's anxiety to get rid of the surplus. But whether their ruin was sudden or gradual, the interests that were locked into the American System felt little inclination to sacrifice themselves to appease South Carolina.[19]

In view of these obstacles to the administration's plan, the opportunity to mediate the crisis reverted to Henry Clay. He had slammed the door on the South only months before. Would he a second time? Rumors that he was readying his own tariff reform appeared in the press even before the Verplanck bill was reported. John Randolph, no friend of Clay's, in a speech on the Virginia hustings, interrupted a tirade against the proclamation, raised a finger and emphatically de-

19. Report of the Ways and Means Committee, December 18, 1832, in *Register of Debates*, 22nd Cong., 2nd Sess., pp. 39–41. On McLane, see John Munroe, *Louis McLane: Federalist and Jacksonian* (New Brunswick, 1973), 365–75. On Verplanck, see Robert W. July, *The Essential New Yorker: Gulian Crommelin Verplanck* (Durham, N.C., 1951), and William Cullen Bryant, "A Discourse on the Life, Character, and Writings of Gulian Crommelin Verplanck," in Parke Godwin (ed.), *Prose Writings of William Cullen Bryant* (New York, 1884), I, 394–431. See also the comments in the New York *Evening Post*, December 14, 31, 1832, and January 4, 1833, and the *Enquirer*, November 23, December 18, 1832.

clared, "There is one man, and one man only, who can save the Union—that man is Henry Clay."[20] But Clay was silent. The crushing defeat at the polls—even worse the resounding Jacksonian victory—had left him despondent. "Whether we shall ever see light, and law and liberty again, is very questionable," he wrote gloomily.[21] Toward the end of the campaign he had been deserted by some men who owed their political careers to the American System; and in defeat he found himself toasted as "the setting sun" of his party. For several weeks he trifled with thoughts of retirement, but on the first of December he was again on the road to Washington, this time with hopes of reviving his political fortunes.

What direction that might take was still obscure to him. He was a shrewd politician, however, and it must have occurred to him that the devious southerly course he had shunned in the presidential campaign—one that involved cooperation with Calhoun and the states' rights republicans of the South —remained open, indeed wide open, under the impact of the president's proclamation. Except for Kentucky he had not carried a single state south or west of the Potomac. At the same time he had lost the protectionist states of Pennsylvania and New Jersey. Perhaps he could devise a strategy of conciliation that would win political favor in the South without further eroding his base of support in the North. He agreed with other National Republicans in condemning Jackson's reckless "double game" at the opening of Congress. "One short week produced the message and the proclamation—the former *ultra* on the side of state-rights—the latter *ultra* on the side of nationalism." But unlike doctrinaire nationalists, Webster for instance, Clay could also say that for all the good of the proclamation it was "entirely too ultra" for him. He left his calling card in the Senate, then hurried off to Philadelphia for an ex-

20. Hugh R. Garland, *The Life of John Randolph of Roanoke* (New York, 1859), II, 361–62. For rumors, see *Enquirer*, December 18, 1832, and C. C. Cambreling to Van Buren, December 29, 1832, in Van Buren Papers.
21. Clay to Charles Hammond, November 17, 1832, in Typescripts, Papers of Henry Clay, University of Kentucky.

tended visit, praying "this unfortunate affair" between the
president and "his brother nullifiers" might be settled without him. [22]

Just what happened in Philadelphia to revive Clay's opti-
mism and good humor remains unclear. His political situation
was not an eviable one. With the Verplanck bill the Jacksonians forced him into a corner where he must either acquiesce
in the destruction of the American System or incur the responsibility of disunion. The alternatives were equally fatal.
How could he defeat nullification and at the same time save
the system? Only by resolving a contradiction greater than
that in Jackson's two papers. How could he yield on the tariff
after the strong position he had taken in the previous session? Only, it seemed, at the risk of becoming a political cha-
meleon. And how could he deny the administration its meditated triumph, transferring the accolades of "Savior of the
Union" to himself? Only at the risk of dealing with Calhoun
and going into the market for southern votes. Without at once
seeing his way through these difficulties, Clay decided to challenge the administration at its own game. Talks with political
friends and associates in Philadelphia, the citadel of the American System, convinced him that Jackson was deeply hostile to
the tariff, as to all the companion measures, and would not
rest until he destroyed it. His strength would be greater in the
succeeding Congress, elected under his banner, than in the
retiring one. The danger from the other side, from South Carolina, which he had earlier dismissed, Clay also took seriously. The American System could be saved, he now realized,
only by surrendering it in some part; and such a negotiated
surrender might save the Union as well.

Some such course of reflection during the leisure of the visit to his ailing brother-in-law, former Louisiana senator James
Brown, now retired in Philadelphia, led Clay to form a com-

22. Clay to Francis Brooke, December 12, 1832, in Calvin Colton (ed.), *The
Works of Henry Clay* (New York, 1904), V, 345–46; Clay to James Caldwell, De-
cember 9, 1832, in Bernard Mayo (ed.), "Henry Clay, Patron and Idol of White
Sulphur Springs," *Virginia Magazine of History and Biography*, LV (1947), 310.

promise plan. Brown was a conciliatory influence; so was Josiah S. Johnston, Brown's successor in the Senate and one of Clay's closest political associates, who was also visiting in Philadelphia. It would be a mistake to say that sugar initiated the compromise; but it was from Johnston, Brown, and their friends—sugar planters and their spokesmen—that Clay learned to what lengths men would subordinate their economic interests in order to save the Union. The true initiators, by his own account, were a committee of manufacturers who called on him "and who asked, with anxiety, what was to be done . . . to save the manufacturing interests from the ruin with which they were threatened."[23] Responding to their plea, Clay worked up a plan that would give the Tariff of 1832 a long lease of seven years, to March 3, 1840, after which date only equal *ad valorem* duties would be levied, "and solely for the purpose and with the intent of providing such revenue as may be necessary to an economical expenditure of the Government without regard to the protection or encouragement of any branch of domestic industry whatever."[24] Whether or not this was Clay's exact language, the project plainly offered the surrender of protectionism after seven years of stability and security. The formula traded *time*, which was of first importance to manufacturers, for *principle*, which was of first importance to the South. The pertinence of the distinction to a workable compromise had been enforced upon Clay by his Virginia friend and colleague, John Tyler. Unlike the Verplanck bill, which would preserve the principle but annihilate the

23. On February 25, 1837, in *Register of Debates*, 24th Cong., 2nd Sess., Senate, pp. 968–69; Brown to Clay, November 5, 1832, in James A. Padgett (ed.), "Letters of James Brown to Henry Clay," *Louisiana Historical Quarterly*, XXIV (1941), 1168; Judge Alexander Porter to Josiah S. Johnston, December 6, 20, 26, 1832, January 16, 1833, in Josiah S. Johnston Papers, Historical Society of Pennsylvania. For the Louisiana background, see Joseph G. Tregle, "Louisiana and the Tariff, 1816–1846," *Louisiana Historical Quarterly*, XXV (1942), 24–148.

24. Reported draft of Clay's first project, copy in the Huntington Library. The source and date of this document are unknown, but in light of later revelations by Daniel Webster, it is probably an accurate report of what Clay originally proposed.

manufacturers, Clay's plan would presumably save the manufacturers but abandon the principle.[25] He promptly consulted
with Johnston, who entered heart and soul into the project.
And when he reported it to the Philadelphia manufacturers
who had prompted him to act, they seemed well satisfied,
buoying his confidence in the possibilities of compromise.[26]

The plan met with a different reception in Washington
whither Clay returned at the beginning of the new year. He at
once unfolded it to John W. Davis, Nathan Appleton, and perhaps other leading protectionists in the House, who were gird-
ing for battle against the Verplanck bill. "We had repeated
interviews," Appleton later recalled these events. "The result
was, *from first to last we refused to become parties to the measure.*"[27] Passing through Philadelphia Webster had heard that
Clay had a plan to settle the controversy, but it was first
explained to him by Kentucky congressman Robert Letcher, a
good friend of both the senators, in the capital. Webster was
aghast. It was a flat proposition to abandon protection, in his
opinion. "From that day Daniel Webster set up for himself,"

25. See John Tyler to Governor Floyd, January 10, 1833, in "Original Letters," *William & Mary Quarterly*, 1st Ser., XXI (1913), 8–10.

26. Clay's biographers and others who have written on the Compromise
Tariff have shed little light on the conception of the first plan, in Philadelphia.
Epes Sargent, *Henry Clay* (Auburn, N.Y., 1852), a campaign life, has some in-
formation. Frederick Nussbaum, "The Compromise of 1833," *South Atlantic
Quarterly*, XI (1912), 337–49, notices the plan but does not explain it. Bryant,
"Verplanck," I, 409–10, wrote of the influence of Eliakim Littell, a Philadelphia publisher, on Clay's plan, but there is no evidence to support this. Clay
himself later recalled the Philadelphia meeting with manufacturers, in *Register
of Debates*, 24th Cong., 2nd Sess., Senate, pp. 968–69, and in a speech at Milledgeville, Georgia, March 19, 1844, reported in *Niles' Weekly Register*, April
20, 1844. In a letter to Hiram Ketchum, January 20, 1838, Webster enclosed
what he declared to be a copy of Clay's original plan. See Papers of Daniel
Webster [Microfilm], Dartmouth College. A brief but generally accurate account of Clay's connection with the compromise is in Glyndon G. Van Deu-
sen, *Henry Clay* (Boston, 1937), Chap. 16.

27. In a speech in the House of Representatives, July 5, 1842, reported in
Boston *Courier*, July 9, 1842. See also Nathan Appleton to Abbott Lawrence,
February 15, 1841, in Nathan Appleton Papers, Massachusetts Historical
Society.

it was later said. He became a favorite at the White House, where his zeal for the proclamation was appreciated, and where, some observers suspected, he hoped to displace Van Buren in the political affections of the president.[28] Clay insisted that his plan abandoned protection only in principle, not in fact. For the present it was secure, and nothing could bind Congress seven years hence. The argument might have been expected to appeal to supposedly pragmatic politicians; but though he had some success with manufacturers, Clay found little support in Congress. As a result, he shelved the compromise plan, renewed his threats to leave the opposing parties to "fight it out" for themselves, and fell back into gloomy forebodings for the future.[29] In the House, meanwhile, debate commenced on the Verplank bill. The Jackson party had the votes for speedy passage, but the protariff opposition bottled up the bill in committee of the whole (where the "previous question" motion could not be put). Even if it passed the House, the bill was given no chance of passage in the Senate.[30]

This uncertainty raised apprehensions of outright conflict between the United States and South Carolina, between Jackson and Calhoun and all they represented. In the president's eyes Calhoun had added treason to his other crimes; in the country generally he rapidly became the personification of nullification and disunion. (Yankee farmers, it was said, turned his somber effigy into scarecrows called "Calhouns.") During the past year nullification had become an obsession with him, "an *idee fixe*," according to one Charlestonian who feared it would snuff out an unrivaled capacity for the manage-

28. Webster to Hiram Ketchum, January 18, 1838, in *Writings and Speeches of Daniel Webster*, XVI, 293; Duff Green, in the *United States Telegraph*, July 20, 1836; Richmond *Whig*, February 2, 8, 1832.

29. Clay to Francis Brooke, January 17, 1833, in Colton (ed.), *Works*, V, 347–48.

30. See the *Enquirer*, January 18, 1832. To the assessment of its Washington correspondent, editor Thomas Ritchie added a nose count that showed twenty senators for the bill, twenty against, and eight undecided or unknown. On opposition tactics in regard to the tariff, see William T. Hammet to F. W. White, January 11, 1833, in William Hammet Papers, Virginia Historical Society.

ment of men and affairs. At every stop on his journey north-
ward in December he had lectured to all who would listen. At
Raleigh, where the legislature was in session, he talked all af-
ternoon and evening—no matter that it was Sunday—to a
constantly changing crowd in a large room of the hotel. With
cavernous eyes blazing under a high forehead and lips quiver-
ing with scarcely suppressed emotion, he expounded tireless-
ly the wonders of nullification. At Richmond he abruptly
announced that South Carolina would pay no heed to the Virginia assembly's dramatic offer to mediate the dispute. The
next day, however, as he attended the assembly and talked
with the delegates, he was more cordial.[31] Arriving in Washington, he felt encouraged by the prospect. The scheme of co-
ercion had been abandoned, at least for the present, though
he thought Jackson was still anxious for it.

The prospect changed on January 16. With the concurrence
of the Carolina Unionist leaders, Jackson sent a special message to Congress calling an end to the truce. Seeing no evidence that the refractory state would abandon nullification
before the effective date of the ordinance, February 1, observ-
ing the army raised by the Nullifiers, responding, too, to
Unionist fears of inability to defend the customshouses, Jackson requested additional authority in two areas: first, to facilitate collection of customs, protect the collectors, and evade
state courts—all with a view of preventing resort to force—
and second, to provide for the more effective and expeditious
use of the army, navy, and militia should force become necessary.[32] Calhoun was visibly excited as the clerk read the message. When it was finished he rose and, with all eyes upon
him, in a trembling voice denounced the president for pro-

31. Washington *National Intelligencer*, May 8, 1838; *Writings of Hugh Swinton
Legaré* (Charleston, 1846), I, 217; R. D. W. Connor, "William Gaston, A Southern Federalist . . ." American Antiquarian Society *Proceedings*, New Ser., XLIII
(1934), 439; *Enquirer*, January 3, 5, 1833.

32. Richardson (ed.), *Messages and Papers*, II, 1173–95. William Drayton to
Joel R. Poinsett, January 13, 1833, in Poinsett Papers, Historical Society of
Pennsylvania, shows Jackson's deference to the Unionists.

posing to impose military despotism in South Carolina. In his manner he reminded a Unionist observer of Milton's description of another vice-regent, and notable nullifier: "Vaunting aloud, but racked by deep despair."[33]

To his friends at home Calhoun urged that they stand by their arms, yet give no pretext for the use of force. The Nullifiers, as if in response, suspended the ordinance while continuing the voluntary resistance of the last two months; a little while later, they decided to reconvene the convention several days after the adjournment of Congress.[34] Nullification was beginning to look like a paper tiger. It was losing even the scare value of which Calhoun had made so much. His insistence on peace and moderation and his warnings against secession seemed to leave the state no option but submission should Congress fail to enact meaningful tariff reform. And in January Calhoun had no reason to believe that it would.

On the other hand, Jackson's request for additional powers —the Force Bill as it came to be called—would surely pass. It was a simple matter of patriotism with many congressmen, regardless of constitutional scruples. Jackson's position throughout the crisis was less clear, more ambiguous, than it seemed. He wanted, it appeared, to put down protectionism and nullification at the same time. But he made no efforts on behalf of the Verplanck bill. After the January message, certainly, he wanted nothing to divert attention from his plan to overawe South Carolina with a show of force. The message, in fact, was generally understood to have blasted the hope of tariff reform; it was even suspected that this was its real purpose.[35] A Virginian who talked to the president early in February said he preferred to postpone the tariff to the new Con-

33. Charleston *Courier*, January 24, 1833; [?] to Joel R. Poinsett, January 23, 18[33], in Poinsett Papers.

34. Calhoun to Armisted Burt, January 16, to James Hamilton, Jr., January 16, to William C. Preston, February 3, 1833, in Clyde N. Wilson (ed.), *The Papers of John C. Calhoun*, (Columbia, S.C., 1977), XII, 15–16, 37–38.

35. See, for example, the Washington letter in the Richmond *Whig*, January 22, 1833.

gress, after he had triumphed over nullification and humbled his adversaries. Webster had much the same opinion of the president's motives. Wanting the whole credit of the victory, Jackson was not inclined to share it with those who would resolve the crisis by removing the ground of complaint.[36] He talked a hard line to South Carolina. The state dared not lift a hand against United States authority, he told a Unionist mission in Washington. "Within three weeks, sir, after the first blow is struck, I will place 50,000 troops in your state."[37] And the fury of the "military chieftain" president, with its threats to the state and its leaders, surely aided in the final settlement.

But the administration could not ignore counsels of prudence and forbearance aimed at keeping the peace within the Democratic party. Every time he raised his voice Jackson widened the split caused by the nationalist doctrines of the proclamation. The administration's own mouthpiece, the *Globe*, obviously embarrassed by the proclamation, tried to explain away its supposed errors. In the circle around the president these were conveniently blamed on Secretary of State Edward Livingston, who had drafted the state paper. The extent of Jackson's complicity in Livingston's errors was confused. Reports that he was unhappy with parts of the proclamation, approved in the heat of the moment, were never substantiated.[38] The Force Bill message, of course, proceeded from these alleged errors. When the redoubtable editor of the Richmond *Enquirer* learned of Jackson's intention to request additional powers from Congress, he bluntly told Senator William C. Rives it would prostrate both president and party in Vir-

36. William T. Hammet to F. W. White, February 4, 1833, in Hammet Papers, Virginia Historical Society; Webster to William Sullivan, January 3, 1833, in Fletcher Webster, ed., *Private Correspondence* (Boston, 1857), I, 528–29; Webster to Joseph Hopkinson, February 7, 1833, in Webster Papers.

37. *Whig*, October 10, 1844, quoting James B. Rhett.

38. The best account of the drafting is Charles H. Hunt, *Life of Edward Livingston* (New York, 1864), 371–81. But see the recollection of John C. Rives, in Frederic Hudson, *Journalism in the United States, from 1690 to 1872* (New York, 1873), 249.

ginia. "Even his best friends would condemn it. . . . The Jackson party in this quarter would in all probability, be shriveled to pieces."[39]

Thomas Ritchie was an alarmist, but the message certainly widened the breach with Virginia states' rightists. The Virginians found support in Albany, where Van Buren and the party faithful in the legislature rejected resolutions endorsing the principles of the proclamation and instead adopted a substitute written by Van Buren that vindicated states' rights principles—the Jeffersonian "Doctrines of '98"—from the heresy of nullification.[40] The New York leader opposed the president's request for additional powers. Even after the Force Bill was reported, it remained a matter of deep regret, if not of opposition, among his followers in Congress. As one of them wrote to Van Buren in Albany: "We are now in the awkward predicament of having the leading measure of the administration ardently supported by the bitterest enemies of the President—ultra federalists—and ultra tariffites who would delight to see the North and South arrayed against each other, while it is now probable that the whole South . . . will go against it." And what was the cause of this predicament? The errors of the proclamation.[41]

The great debate on the Force Bill occurred in the Senate. It was an administration measure, but to the chagrin of party leaders like Van Buren and Ritchie its foremost champion was Daniel Webster. Reminiscing in later years, he boasted that Jackson came to him after the crisis passed, clasped his hand, and declared, "If you and your Northern friends had not come to us when you did, Calhoun and his party would have crushed me and the Constitution."[42] Nullification was a for-

39. Ritchie to Rives, January 6, 1833, in William Cabell Rives Papers, Library of Congress.

40. See *Autobiography of Martin Van Buren*, 550–52; and F. W. Seward (ed.), *The Autobiography of William H. Seward* (New York, 1877), I, 228–29.

41. C. C. Cambreling to Van Buren, February 5, 1833, in Van Buren Papers.

42. Speech at Patchogue, New York, September 22, 1840, *Writings and Speeches of Daniel Webster*, XIII, 117.

tunate event for the Massachusetts senator. The president's proclamation, followed by the special message, opened intriguing possibilities of rapprochement with Jackson and party realignment on the patriotic issue of the Union and the Constitution. When the Judiciary Committee was laboring painfully over the message, Jackson sent Livingston to Webster's lodgings, where he was confined by illness, to plead for the senator's aid and assistance. He roused himself and helped the committee frame an acceptable bill. Reported by the chairman, William Wilkins, it was promptly dubbed "Wilkins' alias Webster's bill." Webster volunteered to act the part of the administration's Cicero in the ensuing debate. As it went forward, however, he held back, seemingly reluctant to risk himself in the forensic encounter with Calhoun that the whole nation breathlessly awaited. Webster had already won his laurels in defense of the Union; he could scarcely hope to embellish them.[43] Calhoun, too, played a waiting game, wishing to reserve to himself the opportunity of making reply. Considering the proposed bill as, in fact, a bill "to make war on a sovereign state," he introduced a series of resolutions designed to bring it to the test of the states' rights principles. Calhoun's whole tendency, in keeping with his reputation as a "metaphysical" politician, was to approach issues theoretically rather than practically and to seek resolution the hard way, on principles, rather than by balancing interests and accommodation to circumstances. But the Senate would not indulge him and proceeded to debate the case of the United States versus South Carolina without clearly determining the principles at issue.

After a week of debate Webster decided it was time "to put the saddle on the right horse." The bill, in other words, was not *his* bill, not the bill of an old Federalist, not the opposition's bill, but the president's own bill, "fragrant of no flower except the same which perfumes the message."[44] This was

43. See the detailed account in Norman D. Brown, *Daniel Webster and the Politics of Availability* (Athens, Ga., 1969), Chaps. 2 and 3.

44. Webster to Joseph Hopkinson, February 9, 1833, in Webster Papers; *Register of Debates*, 22nd Cong., 2nd Sess., Senate, pp. 410–13.

gall to many Democrats who, while they felt compelled to vote for the "bloody bill," wished to place the onus on Webster and his friends. On February 15 Calhoun delivered his major speech against the measure. It was, in fact, his first major speech in sixteen years, which added to the drama of the occasion. Despite a snowstorm, carriages lined the streets before the Capitol, discharging eager auditors who filled the gallery and lobbies of the Senate. The fifty-year-old Calhoun was a gaunt figure, with a pale face, deep sunken eyes, and dark hair that stood straight up from his head. Never an imposing speaker but almost always a penetrating one, he seemed on this day to be consumed by his feelings. He paced back and forth like a caged lion, spoke with dizzy rapidity, gradually sank under the weight of his subject, and after exhausting his voice in vehement denunciation finally became inaudible. Here was the Calhoun that Clay later caricatured as "tall, careworn, with furrowed brow, haggard, and intensely gazing, looking as if he were dissecting the last abstraction which sprang from the metaphysician's brain, and muttering to himself in half-uttered tones, 'This is indeed a real crisis!'"[45]

The crisis, for Calhoun, was between *power* and *liberty*, and there had never been a greater one, even as far back as the Battle of Marathon. The "last abstraction" was *sovereignty*, which in his theory resided in the people of the separate states, not in the aggregate "one people" of the proclamation, and was in its very nature indivisible. The sovereign remedy, nullification, was also an abstraction, involving as it did the political impossibility of a state being in and out of the Union at the same time; and so was Calhoun's bête noir, the tyranny of the majority, which postulated one uniform majority bloc in constant conflict with a uniform minority bloc, though neither existed in reality nor, indeed, in the basic Madisonian

45. *Congressional Globe*, 27th Cong., 1st Sess., 344. For eyewitness accounts of Calhoun's speech, see Henry Moore to Charles Sumner, March 3, 1833, in Typescripts, Papers of John C. Calhoun, University of South Carolina; Henry Barnard, "The South Atlantic States in 1833, As Seen by a New Englander," *Maryland Historical Magazine*, XIII (1918), 283, 308; Charleston *Courier*, February 24, 1833.

theory of American politics, which postulated a variable and pluralistic majority. Even Calhoun's historical analogies, the Roman tribunate as a case of the concurrent majority, for instance, were mistaken.[46]

Webster was on his feet as soon as Calhoun finished. Thinking with most observers that the senator had failed badly, Webster was scornful of his argument. The notes for his reply show careful preparation as well as new research, beyond that of 1830, into the making of the Constitution. "I think *I begin* to understand the Constitution," he wrote to a lawyer friend.[47] With his massive head—the wonder of phrenologists—dark and solemn face, "the dull black eyes under their precipice of brows, like dull anthracite furnaces needing only to be blown," in Carlyle's remarkable description, Webster was a majestic figure befitting his oratorical fame as the Godlike.[48] He and Calhoun, both supposed to be intellectual giants, were often compared: one empirical, the other dialectical in method; one rhetorically spacious, the other austere; one more luminous, the other more penetrating in his views. Webster had the ability of the greatest orators to sweep the fog from even the most bewildering subject and place it in a blaze of sunlight. "He seizes his subject, turns it to the light, and however difficult, soon makes it familiar; however intricate, plain; and with a sort of supernatural power, he possesses his hearers, and controls their opinions."[49] Now, in replying to Calhoun, Webster said nothing new, though some thought he expressed his nationalist doctrine with greater force and clarity than before. The Constitution, he argued, was not a compact among sovereign states but a perpetual union of one people. The issue was whether the Founding Fathers had created a league of states or a national government based on the will of the people. For Webster this was a profoundly patriotic matter, involving not only the question of constitutional powers but

46. Crallé (ed.), *Works*, II, 197–262.
47. Webster to Joseph Hopkinson, February 15, 1833, in Webster Papers.
48. Quoted in Irving H. Bartlett, *Daniel Webster* (New York, 1978), 158.
49. "Ignatius Loyola Robertson" [Samuel Knapp], in *National Intelligencer*, July 17, 1830.

also the legacy of liberty and union received by his generation as a sacred trust from the fathers.[50] The speech was a thrilling performance. "He ground the whole argument of Calhoun to powder," a visitor wrote from the gallery.[51] Few cared to dispute this judgment.

Calhoun, who grimly took notes throughout the two and one-quarter hours, did not reply until ten days later. By then the Senate had passed the Force Bill, with only a handful of absentees and a single dissenting vote. Nevertheless, the senator called his resolutions from the table and made them the basis of a reply to Webster. In an excess of zeal the gentleman from Massachusetts had made the mistake of conceding that if the Constitution could be shown to be a *compact*, the ultimate right of the states, as the contracting parties, to interpret it would necessarily follow. Grant Calhoun the rock—more accurately, his metaphysical premise—and he could build his church. "He spoke absolutely in axioms," it was said.[52] Webster made only a hurried and faltering response. "Aye, he's dead! he's dead, sir!" John Randolph reportedly muttered. "He has been dead an hour ago. I saw him dying muscle by muscle."[53] Webster's northern friends were dismayed that their Hercules had permitted his Antaeus to recover from prostration, rise up, and revenge himself. "Instead of raising him high in the air with a fatal grasp," John Quincy Adams wrote, "Hercules suffered him to march off with a shout of Io Paean."[54]

While the passage of the Force Bill was all but assured from

---

50. *Writings and Speeches of Daniel Webster*, VI, 181–238. Webster spoke eloquently on the Union as a trust in his Worcester speech, *ibid.*, II, 127. See also Paul C. Nagel, *One Nation Indivisible: The Union in American Thought* (New York, 1964).

51. Barnard, "The South Atlantic States," 306; Charleston *Courier*, February 24, 1833.

52. Alexandria *Gazette*, quoted in the Pendleton (S.C.) *Messenger*, March 6, 1833; Charleston *Mercury*, March 6, 1833.

53. *Jeffersonian and Virginia Times*, quoted in *Messenger*, March 6, 1833. This varies somewhat from the more common version of the anecdote, as in Charles M. Wiltse, *John C. Calhoun, Nullifier, 1829–1839* (Indianapolis, 1949), 194.

54. Adams to Robert Walsh, March 29, 1833, in the Adams Papers.

the start, the prospect of tariff reform, never bright, had become desperate by February. The opposition of manufacturing interests was intense. Resolutions of the Massachusetts
legislature instructed the state's senators to vote against the
Verplank bill and declared further that, if adopted, it would
be such a gross abuse of power "as would justify the States
and citizens aggrieved by it, in any measure which they might
think proper to adopt for the purpose of obtaining redress."[55]
(The same legislature, by a similarly overwhelming vote, reelected Daniel Webster to the Senate.) Obviously, the danger
to the Union was not all on one side. In the House the opponents were slowly talking the bill to death. One after another,
day after day, they occupied the floor, while the bill's friends,
with only an occasional interruption, remained silent in the
delusory hope of facilitating passage. "Nothing but a miracle
can save us," wrote one of Van Buren's lieutenants; another
traced the mischief to the rule-or-ruin domination of Congress
by Clay, Calhoun, and Webster. The New York Democrats
were confused and divided. Some advocated tariff reform but
were cool to "the war bill"; others advocated force but stood by
protection; and in the absence of clear signals from Van Buren
at Albany, he and his friends were suspected of opposing a
settlement on any basis.[56] On February 8 the House, still in
committee of the whole, began the tedious process of amend-
ing the Verplanck bill, though everyone realized it stood no
chance of passage and the leadership searched frantically for a
way out of the parliamentary impasse.

It was at this time that Henry Clay astounded Congress and
the nation by the introduction of his plan to compromise the
tariff. Two or three weeks earlier he had quietly resumed the
effort, never completely abandoned, to arrange a settlement.
Rumors of bargain and coalition between Clay and Calhoun
were unceasing, but they became more creditable as the session advanced. McLane detected signs of an understanding

55. *Evening Post*, January 26, 1833.
56. Michael Hoffman to A. C. Flagg, February 4, 1833, in Flagg Papers [Mi-
crofilm], New York Public Library; Silas Wright to N. P. Tallmadge, January
27, 1833, in Tallmadge Papers [Microfilm], Wisconsin Historical Society.

near the end of January; the fact of such an understanding, growing out of conversations between the two senators,
was prominently reported in a Philadelphia newspaper on
the twenty-fifth.[57] Clay had been burned once by the charge
of "corrupt bargain," and still carried the scar. Why would he
choose to expose himself a second time? Pondering this question some men gave him the benefit of lofty motives, of disin-
terested patriotism, while others suspected him of indulging
a natural political aptitude for self-serving games of bargain
and intrigue. In the case at hand he did not need to do anything, of course; and after the cold reception of his first plan
to reduce the tariff, he had petulantly retired to a neutral corner. Any conciliatory move, clearly, would disgust many of
his friends. And what could he possibly hope to gain from
enemies as implacable as Calhoun, McDuffie, and the whole
crew of Nullifiers? He and Calhoun had once been allies, war
hawks together in 1812 and nationalists united on the Madisonian platform after the Peace of Ghent; but since 1824 the political rift between them had widened until they now agreed
upon nothing except opposition to Jackson and Van Buren.

Clay could not, however, overcome his anxiety for the fate
of the American System at the hands of the administration.
There might be little danger of destructive legislation at this
short session, but it seemed certain in the new Congress.
"Jackson has decreed its subversion, and his partizans follow
him wherever he goes," Clay wrote pessimistically. "He has
marked out two victims, South Carolina, and the Tariff, and
the only question with him is which shall first be immolated."[58]
To salvage what he could of the American System, to keep the
policy in friendly hands even in dissolution, became an important object. Clay was also concerned to deny the administration the glory of peacemaking, on one side, or of Bloody

---

57. Clay to Alexander Porter, January 11, 1833, in Typescripts, Papers of
Henry Clay, University of Kentucky; McLane to Van Buren, January 23, 1833,
in Van Buren Papers; Webster to Edward Everett, January 26, 1833, in Webster
Papers; and for the editorial in the Pennsylvania *Inquirer*, see Charles M.
Wiltse (ed.), *The Papers of Daniel Webster* (Hanover, N.H., 1974), III, 209n.

58. Clay to Alexander Porter, January 29, 1833, in Typescripts, Clay Papers.

Bill triumph on the other. The vision of President Jackson marching into South Carolina at the head of a Union column filled him with horror. Expecting the state to submit, Webster and others like him never took this danger seriously; and if it occurred, if the Union was brought to the test of force, they would be more elated than dismayed. But Clay feared conflict, especially if the Force Bill was enacted and the Nullifiers returned home without even the semblance of tariff reform. Clay saw, too, that the Carolinians "were extremely unwilling that Jackson should have any credit in the adjustment of the controversy, and to prevent it were disposed to agree to much better terms for the manufacturers, if the measure originated with any other."[59] He availed himself of this disposition. No one but Clay possessed the influence or the prestige to do so.

Early in February Clay matured his second, or revised, compromise plan. Like the first, it called for the reduction of protective duties to the revenue level, which was targeted at 20 percent *ad valorem*, but instead of a grace period of seven years, the new plan proposed to effect the change gradually over nine and one-half years. The idea of a gradual withdrawal of protection, in the interest of manufacturers, had been talked about for several years. John Tyler, his former colleague Littleton Tazewell, and Governor Hayne, when in the Senate, had all advocated it in one form or another. Coming from southerners who held the protective tariff to be unconstitutional, it was meant as a concession. Now Clay accepted it as such, as the basis of compromise, though a decade-long concession of protection was more than the southerners had bargained for. Every great compromise involves mutual concessions and the adjustment of opposing interests between the parties. The achievement of this is high politics as far as politics can be considered the art of the possible. It requires, of course, a belief in the positive values of compromise, as something potentially desirable as well as necessary, progressive instead of regressive, moral instead of immoral, which Clay possessed

59. Clay to Thomas Speed, June 9, 1833, *ibid.*

beyond any of his peers. With him, compromise, far from being the subversion of principle, might be its fulfillment.

Clay first worked through the details of his plan with protectionist friends in the Senate. Eleven or twelve of them were summoned to an evening meeting at his lodgings, which was quickly followed by another. No roster has survived, but the senators in attendance probably included Peleg Sprague, of Maine; Samuel Bell, of New Hampshire; Samuel Foot, of Connecticut; Theodore Frelinghuysen, of New Jersey, John M. Clayton, of Delaware; Josiah Johnston, of Louisiana; and Thomas Ewing, of Ohio.[60] Webster attended the first of the meetings but not the second. He tried to prepare his friends for "an explosion" in the Senate. "It is understood Mr. Clay will agree to almost anything, in order to settle the question, save the Nullifiers, and obtain the credit of pacification," he wrote. Professing to know nothing of the particulars, he conveyed the impression he had not even been consulted. "*I am not party to the protocols*," he emphasized.[61] This was a half-truth at best; if Webster was ignorant of "the protocols," it was because he chose to be. At the second meeting, on February 5, Clay interrogated each of the senators until, finally, all but one pledged himself to support the proposed plan. He also consulted prominent manufacturers like E. I. DuPont (one of the earlier Philadelphia group), who converged on the capital in February, John Sergeant, of Pennsylvania (his running mate in the recent election), and leading protectionists in the House.[62]

Proceeding in this way Clay soon found enough encouragement among his friends to warrant negotiation with his enemies. Using the affable Kentuckian Robert Letcher as

60. References to the meetings may be found in Clay to Webster, February 5, 1833, and to Nicholas Biddle, April 10, 1833, in Webster Papers.
61. Webster to Joseph Hopkinson, February 9, 1833, and to Nathan Appleton, February 9, 1833, *ibid.*
62. Clay to Josiah Johnston, March 15, 1833, in Typescripts, Clay Papers; Sargent, *Clay*, 141; Bessie G. DuPont (ed. and trans.), *Life of Eleuthère Irénée DuPont from Contemporary Correspondence* (Newark, Del., 1923); Michael Hoffman to A. C. Flagg, February 7, 8, 1833, in Flagg Papers [Microfilm].

an intermediary, Clay communicated his plan to Calhoun. By this time Calhoun was anxiously seeking an apology for retreat from nullification and seems only to have been waiting for the Pacificator to make his move. Letcher, finding him receptive, apparently arranged an interview between the two leaders at Clay's boardinghouse. Thomas Hart Benton later characterized this meeting as "cold, distant and civil" and ending "without result."[63] Calhoun, though amenable to a gradual withdrawal of protection, called for a shorter term of four or five years and a sharper reduction to 15 percent. And that might have been the end of it but for the perseverance of Johnston and Letcher. The Louisiana senator, on Benton's testimony, relayed to Letcher the latest report of the president's rage against South Carolina, including the threat to hang the chief of the Nullifiers as a traitor; and Letcher, in the middle of the night, went to Calhoun's lodgings and startled him out of bed with this horror, thereby securing his consent to the compromise as the price of escaping the gallows. The story is almost too good to be true.[64] At any rate, the conversations were resumed, apparently on Calhoun's initiative. Professing to be the disciple of absolute truths, Calhoun, unlike Clay, could compromise only at grave risk. "Expediency, conces-

63. Thomas Hart Benton, *Thirty Years' View* (New York, 1854), I, 242. Benton is the principal source for the meeting. Clay later denied it ever occurred: *Register of Debates*, 24th Cong., 2nd Sess., Senate, pp. 968–70.

64. During the Civil War Benjamin O. Tayloe, a longtime resident of Lafayette Square, in Washington, recorded Letcher's recollection, in 1857, of an interview he had with the president, at Clay's behest, in order to restrain his vengeance against the South Carolinian:

JACKSON: "Do you not think Calhoun deserves the gallows?"

LETCHER: "I will not deny that; but the question is the effect upon the country, *your* administration, and *your* reputation."

JACKSON: "As Calhoun deserves punishment, what would you have me do?"

LETCHER: "Postpone your decision. Take time for cool reflections, and consult your friends in the Cabinet individually, and your true friends, in whom you confide."

After further discussion Jackson agreed to the advice and Letcher, according to this report, set out to save the peace together with Calhoun's neck. [Winslow M. Watson], *In Memorium: Benjamin Ogle Tayloe* (Philadelphia, 1872), 102–103.

sion, compromise! Away with such weakness and folly!" But while this was the image he sought to maintain, he was never enslaved by it. In the present crisis he realized that compromise offered the only way out and, further, that if he did not strike a bargain with Clay he would be unable to bargain at all. [65]

Despite the interviews, the consultations, and the rumors in the press, Clay's speech on February 12 marked such an abrupt change of public position that it took nearly everyone by surprise. The bill, which he requested leave to introduce, provided for the continuation of the Tariff of 1832 with major modifications. From January 1, 1834 (it was first proposed on October 1, 1833), all duties over 20 percent would be reduced in biennial installments of one-tenth, with one-half the residue—six-tenths of the whole—taken off on January 1, 1842, and the other half on July 1 of the same year. After that, as stipulated by the third section, "such duties shall be laid for the purpose of raising such revenue as may be necessary to an economical administration of the Government." Clay was careful to explain that Congress could raise, or lower, the rate of duties from the 20 percent level if "the exigencies of the country" required. Although the bill contained no disavowal of protection, it seemed to establish a uniform 20 percent *ad valorem* rate, without discriminating or specific duties, as the norm of a revenue tariff. Such a tariff would afford only incidental protection. Other sections of the bill provided for additions to the duty-free list, for future abolition of the credit system, and for restoration of the high protective duty on cheap woolens. (The nominal 5 percent duty on these articles had been offered as a concession to the South in 1832; but planters spurned it as being of no consequence, saying they did not

65. Crallé (ed.), *Works*, III, 190; Benton, *Thirty Years' View*, I, 342–43. Benton's account of this and other aspects of the compromise is often in error, though useful and interesting. George T. Curtis, *Life of Daniel Webster* (New York, 1870), I, 444, asserts on the authority of John J. Crittenden that Calhoun sought out Clay. Wiltse, *Calhoun, Nullifier*, Chap. 14, places the meeting and the agreement in January, which does not seem possible. John Tyler claimed that he brought Clay and Calhoun together, in Lyon G. Tyler, *Letters and Times of the Tylers* (Richmond, 1884), I, 456–59, 467.

use such low-grade cloth even for their slaves, so Clay proposed to restore the old protective duty.)[66]

One of Clay's grand purposes, of course, was to preserve
the Union. If his tariff compromise won southern support, it
would accomplish this without the fulminations of the Force
Bill. The objection to concession in the face of nullification,
earlier so much felt, Clay now argued, had been largely overcome by the civil manner of South Carolina's resistance. Nullification was not only suspended, it was dead, the victim of
a far mightier sovereign, public opinion, he declared. There
was no immediate threat to the Union from South Carolina;
the threat came, rather, from the Jackson administration. A
conciliatory adjustment of the tariff was necessary to restore
the harmony and ensure the permanence of the Union. All
this was said in keeping with the senator's role as peacemaker
and, of course, to combat the charge of receding before the
menace of nullification.

But Clay was more attentive to his other role, as the great
advocate of the American System, and to his second purpose,
the preservation of the protective tariff. It stood in imminent
danger; its overthrow would be as disastrous for America,
he said, as the revocation of the Edict of Nantes had been for
the French nation. Anticipating the objection that his bill
would, in fact, overthrow his system, Clay vigorously denied
it. "What is the principle which has always been contended
for? . . . After the accumulation of capital and skill, the manufacturer will stand alone, unaided by the government, in
competition with the imported articles from any quarter. Now
give us time; cease all fluctuations and agitations, for nine
years to come, we can safely leave to posterity to provide for
the rest." During those years the tariff would be not only secure, with only the most gradual remissions, but separated
from politics so that businessmen could count on the stability
of the law "without every thing staked on the issue of elections, as it were on the hazards of the die." Conceding that no

66. The original bill (S115), in Clay's hand, with related papers, is in the
Senate Records, RG 46, National Archives.

statute could permanently bind Congress, he thought that
the circumstances giving rise to this act would elevate it above
ordinary legislation to the status of a "treaty of peace and am-
ity." The act would be a true compromise, a mutual accommodation between the long-contending interests of manufacturers and planters, in which the former received the security
of intermediate protection, the latter the promise of ultimate
reduction to the revenue standard. Each side surrendered a
little of its particular interest for the transcendent interest
of both in the American Union. "The distribution," Clay concluded on a philosophical note, "is founded on that great
principle of compromise and concession which lies at the bot-
tom of our institutions."[67]

When Clay sat down, several administration senators rose
to denounce his motion. The audacity of the man whose system of politics was responsible for the crisis, in their view,
suddenly throwing his solution into Congress when only
eighteen days remained of the session! Mahlon Dickerson, the
New Jersey protectionist, said he would not vote for leave. The
high point of the day's proceedings came when Calhoun rose
to declare his approval of the principles of the bill. He had
never, he said, wished to prostrate the manufacturers by too
sudden a withdrawal of protection, as proposed in the House
bill, and had always looked to a just and enduring settlement.
"Such was the clapping and thundering applause when Calhoun sat down that the Chair ordered the galleries to be
cleared," wrote one observer. "The sensation was indescrib-
able."[68] In the House, meanwhile, the bargain between Clay
and Calhoun was sealed by the election of Gales and Seaton,
publishers of the opposition *National Intelligencer*, printers of
the House. A week later Calhoun's friend and advocate, Duff
Green, would be elected printer of the Senate. These were lu-
crative positions, and more importantly they were positions
of political influence. The division of the spoils between Clay

67. Daniel Mallory (ed.), *The Speeches of Henry Clay* (New York, 1843), II,
106–21.
68. William T. Hammet to F. W. White, February 12, 1833, in Hammet
Papers.

and Calhoun in these elections showed the strength of "the new partnership," not only positively but negatively by the humiliating exclusion of Blair and Rives, the administration's publishers.[69]

Leave was granted; the bill was immediately taken up and on the next day referred to a select committee under Clay's chairmanship. This was auspicious, for either of the appropriate standing committees, Manufactures or Finance, would have been hostile. Among protectionists in and out of Congress the bill found more critics than defenders. It came "like a crash of thunder in the winter season," said *Niles' Weekly Register*, long a powerful voice of the American System. While circumstances may have required some change, it was scarcely creditable that the author of the system proposed to save it by abandoning the very principles of discrimination and protection on which it was built. The bill was offered as a compromise, yet all the sacrifice—the doom of northern prosperity—was on one side, according to the Boston *Courier*. As to saving the Union, supposing the Union was in danger, which Clay himself doubted, it could be asked: "But is the Union to be forever a matter of compromise? Is there nothing certain, stable or efficient in the original compromise, which is embodied in the Constitution? . . . And must the terms of the Union be forever the subject of litigation, temporizing and compromise?"[70]

To ask such questions was to answer them. It was easy, of course, to read sinister motives into Clay's conduct: to say that he had crossed the Potomac and offered to rescue the Nullifiers in exchange for the vote of the South for president. But most protectionists, though they disagreed with Clay, did not question the patriotism of his motives. Most Jacksonians did little else. The compromise bill was a clever maneuver by Clay to escape political annihilation, they said. "Will you gen-

69. On the significance of the election, see especially the comments of "The Spy in Washington" [Mathew Davis], in the Albany *Evening Journal*, February 17 and 28, 1833.

70. *Niles' Weekly Register*, February 16, 1833; Boston *Courier*, February 20, 1833.

tlemen of the Senate allow him this new *feather* in his cap! Will you let the vanquished dictate to the victors?" an astonished Virginian asked. And, what was worse, the price Clay exacted for the revival of his own political fortunes was the revival of Calhoun's! "It is all got up by the black hearted revenge of Clay and Calhoun towards Jackson."[71] To which Jackson inclined to agree.

The select committee appointed by the president *pro tem* of the Senate, Hugh Lawson White, of Tennessee, consisted of Clay, Felix Grundy, also of Tennessee, George M. Dallas, of Pennsylvania, Webster, Calhoun, Rives, and Clayton. Several years later, long after he broke with the administration, White testified that Jackson intervened to prevent Clayton's appointment, mainly, it seems, to gratify his Delaware rival, Louis McLane, whose tariff bill languished in the House.[72] Whatever the truth of the story—Jackson flatly denied it—the administration was confounded and dismayed by Clay's bill. The select committee met more or less daily while the Force Bill was before the Senate. Webster and Dallas were unremittingly opposed. The Bay State senator had stated his position in a series of resolutions presented to the House but passed over when Clay's bill was referred. The tariff should be gradually reduced to the revenue level, but selectively rather than uniformly, thereby preserving the principles of discrimination and protection. In this regard the administration bill was preferable to Clay's, which Webster thought forever and completely surrendered protectionism. Dallas epitomized the dilemma of the Pennsylvania Democrat, but regarding Clay's bill as "a mere political maneuver," he easily rejected it. Grundy, as a loyal Jacksonian, favored the administration bill yet

71. C. W. Gooch to William C. Rives, February 16, 1833, in Rives Papers. See also Edward Everett to A. H. Everett, March 13, 1833, in Papers of Edward Everett [Microfilm], Massachusetts Historical Society; John Quincy Adams to C. F. Adams, March 26, 1833, in Adams Papers. For administration views, see *Evening Post*, February 14, 1833, and the Washington *Globe*, February 16, 1833.
72. N. N. Scott, *A Memoir of Hugh Lawson White* (Philadelphia, 1856), 239–40, 299–300, 326; Calvin Colton, *Life and Times of Henry Clay* (New York, 1846), II, 470; Lexington, Ky., *Gazette*, June 29, 1837.

went along with Clay's compromise effort. Calhoun and Rives were for it, of course. And so was Clayton, an adoring protégé of the Kentuckian, though he wanted protectionist amendments. There were, then, more than enough votes in committee to report the bill. But were there enough votes to pass it in the Senate, or in the House? This was the question that troubled Clay, that led Webster to portray him as "half-sick" of his own measure, and that enabled Clayton to raise the stakes for Calhoun.[73]

Just how it happened that Clayton took the fate of the measure into his own hands remains uncertain. According to one contemporary account, the several members (six senators and one representative) of his Capitol Hill mess, every one a protectionist, demanded one or more crucial amendments as the price of their support for the compromise. Clayton agreed and became the spokesman for this protectionist bloc. According to another account, Clayton assembled the manufacturers who had gathered in Washington, and acknowledging that nothing could be done without their consent, obtained their terms for acceptance of the compromise.[74]

However it came about, Clayton, acting as the broker for manufacturing interests, demanded that the tariff be levied on the *home valuation* from the ultimate date of Clay's bill. Foreign valuations were notoriously understated on the invoices; the vice of "fraudulent invoices" had always been a protectionist argument for specific rather than *ad valorem* duties. If the country was going exclusively to the latter, home valuation would add the equivalent of five to ten points to the tariff duties and thus afford some compensation for the loss of protection. The same principle, it may be recalled, had been incorporated by Clay into his bill at the previous session. Clayton

73. George M. Dallas to Henry D. Gilpin, February 19, 1833, in Dallas Papers, Historical Society of Pennsylvania; Webster to Joseph Hopkinson, February 15, and to Nathan Appleton, February 17, 1833, in Webster Papers.

74. Sargent, *Clay*, 142–43; Benton, *Thirty Years' View*, I, 343. On February 13 Clayton wrote to E. I. DuPont (the letter is in the Eleutherian Mills Historical Library) asking advice on the course he should pursue and urging DuPont to return to Washington.

and his friends did not trouble themselves to explain how varying valuations at ports of entry from Wiscasset to New Orleans could be squared with the constitutional requirement of uniformity of duties. The Delaware senator was also deeply interested in Clay's Land Bill, both because it would distribute revenue to the states for internal improvements and because it would increase the federal Treasury's dependence on the tariff.[75] (Theoretically, of course, the government could resort to direct taxation, freeing it of dependence on the tariff; but no one, not even Nullifiers and free traders, advocated this.) Home valuation, together with distribution of the proceeds from the sale of public lands, would go a long way toward securing adequate protection after July 1, 1842. The committee heard Clayton's demand, and firmly rejected it. Calhoun and the other southern members considered home valuation outrageously protectionist as well as unconstitutional. Webster and Dallas considered it impracticable, possibly unconstitutional; more importantly, they opposed the spirit and principle of the compromise itself. The committee finally voted four (Clay, Calhoun, Grundy, and Rives) to three (Webster, Dallas, and Clayton) to report Clay's bill to the Senate with only one minor amendment that permitted discriminatory duties below the level of 20 percent from the ultimate date.

Surprisingly, when Clay opened the Senate debate on February 21, he offered the home valuation amendment as his own. By throwing his weight behind this artifice he hoped to regain the affections of alienated protectionists, though he thereby jeopardized his arrangement with Calhoun, who at once declared he would vote against the bill if the amendment was adopted. To this decree Clayton coolly responded with his own, making the amendment the necesary condition for his vote. Why was the Senate occupied with this business

75. See Clayton's speech of June 15, 1844, in Colton, *Life and Times of Henry Clay*, II, 578*n*; Joseph P. Comegys, *Memoir of John M. Clayton* (Wilmington, 1882); Clayton to DuPont, March 2, 1833, in Eleutherian Mills Historical Library; and draft of a letter [April, 1841] to Nicholas Biddle, in John M. Clayton Papers, Library of Congress. See also, Richard A. Wire, "Young Senator Clayton and the Early Jackson Years," *Delaware History*, XVII (1976), 104–26.

at all? he asked. To conciliate South Carolina. If South Carolina, in the person of John C. Calhoun, spurned the condition of conciliation, the Senate should cease to trouble itself; and he moved to table the bill. The threat, if serious, was not final. Clayton promptly withdrew the motion at the request of a southern senator. Debate continued into the evening hours. Restrained by Clay, the Delaware senator did not renew his killing motion; and rather than risk a nocturnal vote on the amendment, possibly jeopardizing the entire bill, his protectionist friends abruptly moved to adjourn. The motion, which in the circumstances was a test of strength between the two sides, passed 22 to 18. The protectionists voted for it, of course, while two-thirds of the nay votes were from southern senators and included both Clay and Calhoun.[76]

That night the South Carolinian no doubt reflected on his course. He may have met with Clay.[77] The next morning several antitariff senators (not all southerners) suggested that Calhoun and his colleague, Stephen Miller, be allowed to vote against the amendment but for the bill. Clay begged Clayton to accept this face-saving proposition. He refused. The amendment had become a symbol, positively, of the survival of protection and, negatively, of the surrender of nullification. If the South Carolina senators wanted to save their necks from the halter, Clayton said, they must vote for the parts as well as for the whole bill, leaving no grounds for cavil in the future. Miller then came to him, agreeing to vote for the amendment but pleading for Calhoun to be spared this humiliation. Again Clayton refused. Taking out his watch, he threatened to nail the bill to the table unless he had the senator's capitulation in fifteen minutes. Miller returned with it in ten minutes. "Very good, you have saved your necks from a halter," Clayton replied.[78] Before the roll was called, Calhoun rose to say he

76. *Register of Debates*, 22nd Cong., 2nd Sess., Senate, pp. 694–97; Charleston *Courier*, March 1, 1833. See also the account in Nathan Sargent, *Public Men and Events* (Philadelphia, 1875), I, 236–39.

77. This is stated in the Washington *Globe*, February 28, 1833.

78. Sargent, *Public Men*, I, 338–39. (The account errs in referring the dispute to the vote on the bill instead of the amendment.)

would vote for the home valuation amendment, despite strong objections, because the fate of the bill had come to depend on it. The amendment passed 26 to 16. Most of the southern senators voted aye, the main exceptions being administration stalwarts.

The subsequent debate in the Senate featured a dramatic encounter between Clay and Webster. Clay was a great orator, of course, but unlike Webster, who excelled in set pieces, he was most captivating in the spontaneous repartee and flashing verbal swordplay of the legislative forum. Fighting off amendments that would unravel the compromise so laboriously put together, he was characteristically candid and forthright. An amendment to strike the protective duty on cheap woolens, for instance, would remove one of the pillars, for the provision showed on its face that preservation of protection was a leading object. Webster thought that this was another piece of the humbug that composed the bill. He never recovered from the shock of Clay's original draft, in January, which not only looked to equal duties and a revenue standard but, Webster always insisted, expressly disavowed protection or encouragement of domestic industry. He believed this was still the intent and that Clay equivocated to gull the protectionists. "The gentleman from Kentucky supports the bill from one motive, others from another motive. One, because it secures protection; another because it destroys protection." While this condemned the measure to "absurdity" in Webster's eyes, it constituted the bill's great political merit—the genius of the compromise—in Clay's.[79] Webster was troubled by other ambiguities. Some advocates of the bill lauded its immediate benefits, in pacifying the country, and made light of clauses that undertook to control Congress in the future. "They do not halloo till they are out of the wood," Webster remarked. Once the bill became law, he predicted, they would insist that Congress was *pledged* to the attainment of a uniform *ad valorem* tariff for revenue only. "For a poor lease of eight years, we surrender the inheritance," said Webster. And why? Not

79. *Register of Debates*, 22nd Cong., 2nd Sess., Senate, p. 723.

to avoid a civil war, the danger of which had passed, if it had ever existed, but "a law suit . . . a war of processes" between the United States and South Carolina.[80]

Tempers flared when Clay charged that Webster, backed by the administration, stood in the way of "this measure of pacification," yet hurled the thunder of the Force Bill at South Carolina. "Would the Senator from Massachusetts send his bill forth alone without this measure of conciliation?" Clay asked. Webster again denied that it was *his* bill. Some observers thought they then heard Webster throw back the challenge, charging that Clay *dodged* the vote on the Force Bill three days before. In fact, he did not make this charge—one that the Kentuckian would have spiritedly repulsed as an attack on his character—though he may have thought it justified. Clay had been absent the night the Force Bill came to a vote. He apologized the next day, saying he could not endure the atmosphere of the chamber after the lamps were lighted and, with a number of the absentees, had not expected a vote that evening. (He had also been absent two days earlier, at 11:30 P.M., when the bill passed to a final reading.) Clay freely avowed then and later that if he had been present he would have voted for the bill, reluctantly, not from any question of its propriety but from distrust of the administration. There is no reason to doubt him (only one senator, Tyler, actually voted nay on passage); still the circumstances had been wonderfully convenient for him. His extraordinary silence throughout the Force Bill debate testified more eloquently than words or votes that he considered support of the measure incompatible with his role as peacemaker and conciliator of the South.[81]

Without concert or arrangement, but in the nature of the process, it was coming to be recognized on both sides, by

80. "Notes of a Speech on the Compromise Bill," *Writings and Speeches of Daniel Webster*, XIV, 586–87. The speech was never delivered. For the circumstances, and the original publication of the notes, see the New York *American*, January 22, 1840.

81. *Register of Debates*, 22nd Cong., 2nd Sess., Senate, pp. 722–25, 689. On the "dodge" charge, see the encounter between Duff Green and "Vindex"

Clay and Calhoun, in the administration and in Congress, that the Compromise Bill and the Force Bill would pass into law together. The marriage of these opposites was, in fact, the compromise for many men who objected to one or the other bill by itself. It was not an acceptable marriage to Webster and his friends, however, since one bill canceled the other by yielding the very principle of the laws it was intended to enforce. The Compromise Bill took the starch out of the Force Bill and its vaunted vindication of the Union. Clay elaborated on his conception of the relationship between the two measures in the remarkable speech he delivered to close the Senate debate on February 25. Of the Force Bill he said, "I could not vote against the measure; I would not speak in its behalf." Unlike some, he had discovered in this crisis "no new born zeal" for the Jackson administration. The proclamation was not only "ultra" in its doctrines, it was inflammatory in its effects, more likely to provoke than to prevent civil war, he said. The tendency of the Force Bill, even conceding its necessity, was the same. But the Compromise Bill would restore the balance between law and order and peace and concord. "The difference between the friends and the foes of the compromise . . . is that they would, in the enforcement act, send forth alone a flaming sword. We would send out that also, but along with it the olive branch, as a messenger of peace. They cry out, the Law! the law! the law! Power! power! power! We, too, reverence the law, and bow to the supremacy of its obligation; but we are in favor of the law executed in mildness, and of power tempered with mercy." Hurried along by a vivid imagination, he continued. "They would hazard a civil commotion, beginning in South Carolina and extending, God only knows where. While we would vindicate the federal government, we are for peace, if possible, union, and liberty. We want no war, above all, no civil war, no family strife. We want no sacked cities, no desolated fields, no smoking ruins, no

---

[John·M. Clayton] in the *United States Telegraph,* July 20 and 29, 1836. Clay discusses the circumstances in a letter to P. R. Fendall, August 8, 1836, in Typescripts, Clay Papers.

streams of American blood shed by American arms!" Every heart throbbed at this burst of eloquence.[82]

The speech throughout was a lively expression of Clay's genius for accommodation, of his sensitivity to opposing political pressures and his ingenuity in adapting old policies to new conditions. Webster, of course, charged him with surrendering the protective tariff in the face of intimidation, under the influence of panic, to no useful public purpose. Clay was not deterred. He honestly believed it was necessary, not to surrender, but to contract the tariff in order to save it. While Webster was apprehensive of the future, Clay willingly took his chances with it, believing the arts and industries of the country would grow stronger under the compromise. "If they can have what they have never yet enjoyed, some years of repose and tranquility, they will make, silently, more converts to the policy, than would be made during a long period of anxious struggle and boisterous contention. Above all," he continued, "I count upon the good effects resulting from the restoration of the harmony of this divided people, upon their good sense and their love of justice. . . . And how much more estimable will be the system of protection, based on common conviction and common consent, and planted in the bosoms of all, than one wrenched by power from reluctant and protesting weakness?"[83] This was in the spirit of Calhoun's search for union and concurrence in federal policy, but Clay sought to show that the goal could be reached through the give and take of the political process, without the awkward and ruinous contrivance of the theory of nullification.

Even as he was speaking Clay had decided to suspend further action in the Senate and send the bill to the House of Representatives. From the beginning opponents had cited the constitutional mandate that legislation for "raising revenue" originate in the House; and although this was countered by

---

82. *Register of Debates*, 22nd Cong., 2nd Sess., Senate, pp. 137–38. For a slightly varying report, and comment on the audience reaction, see the *Whig*, March 8, 1833.

83. *Register of Debates*, 22nd Cong., 2nd Sess., Senate, pp. 731–32.

the sophistry that the bill looked to lowering, not raising, revenue, the sensible course was to avoid constitutional objections. The decision may also have been part of an informal understanding with Jacksonian leaders, who now agreed to support Clay's measure if it went through side by side with the Force Bill in the House.[84]

The representatives had begun debate on the latter, but still had found no way out of the parliamentary bog into which the Verplanck bill had taken them. Thus it was that on Monday evening, February 25, suddenly and without warning, just as congressmen were putting on their wraps to go home, Clay's Kentucky friend Letcher rose and moved to refer the Compromise Bill immediately to the committee of the whole house with instructions to substitute it for the Verplanck bill. Received as an act of deliverance, the motion was quickly approved; and over feeble cries of protest against legislating with a gun at their heads, the congressmen passed the bill to a third reading, 105 to 71, before going home. The next day northern opponents issued dire prophesies for the fate of the Union from compromise with nullification. Tristam Burges, of Rhode Island, poured forth a doleful lament on the fading "Star in the West." "He has been the hero of our tales. . . . Whose —whose name was ever so frequent in our flaming cups! Who, like him, ever sat with us at the great political table! We have taken salt from the same stand; bread from the same basket . . . and now he has lifted up his heel against us, and we are delivered unto the hands of our adversaries."[85] But after only two or three hours of debate, appropriately closed by McDuffie's avowal of support, the House adopted the Compromise Bill, 119 to 85. As it returned to the Force Bill,

84. Nathan Appleton, in the House of Representatives, July 5, 1842, as reported in the Boston *Courier*, July 9, 1842, said he was informed by James M. Wayne, an administration leader in the House, that a caucus of the party had met on February 24, at which it was decided to support Clay's bill and arrange for its introduction in the House. This is plausible. On the administration's decision to support the bill, see Silas Wright to A. C. Flagg, February 25, 1833, in Flagg Papers.

85. As reported in the *National Intelligencer*, February 26, 1833.

McDuffie wondered at the justice or policy of fettering "the olive branch of peace" with "the sword of blood." But in the eyes of most congressmen one had become essential to the other. The Force Bill passed, March 1, on the vote of 149 to 48. On the same day, the Senate passed the Compromise Bill without further ado, 29 to 16.

Henry Clay called March 1 "perhaps the most important Congressional day that ever occurred." It was a personal triumph, of course—"the most proud and triumphant day of my life," he told a reporter.[86] The House sent the Force Bill, the Senate both the Compromise Bill and the Land Bill to the president for signature. The latter was the almost forgotten child of the session. In December Clay had introduced the same measure the Senate had approved and the House had postponed in the previous session. Once again the Public Lands Committee, with its western outlook, reported its own bill; and, again, the Senate adopted Clay's plan for distribution of the proceeds from the sale of the lands to the twenty-four states. After years of discussion the question was well understood. If the proceeds of the Land Office were excluded from the revenue to support the government, government would be solely dependent on the tariff, which would tend to perpetuate the American System. Logically, then, distribution was part of the compromise, another compensatory device, like home valuation, to keep up the tariff. Some Jacksonians imagined that this measure—"the most mischievous . . . that ever originated in Congress"—was the principal consideration in Clay's bargain with Calhoun. "The Land Bill," one wrote, "is to *reward* Kentucky for sacrificing *the System*."[87]

Actually, however, the Senate acted on distribution before either enforcement or tariff bill came to the floor, even before Clay had any conversations with Calhoun. Distribution did not enter into the negotiations; it was the object of no pledge or promise, and in that sense was no part of the compromise.

86. Clay to James Barbour, March 2, 1833, in Typescripts, Clay Papers; "The Spy in Washington," in *Whig*, March 15, 1833.

87. Michael Hoffman to A. C. Flagg, February 23, 1833; Silas Wright to A. C. Flagg, January 27, 1833, in Flagg Papers.

Calhoun continued to oppose distribution; with most southern and western senators he voted against the bill. Clay, although he underscored the relationship between the two measures, also stated he did not consider them so far united as to
stand or fall together. No significant correlations are revealed
by the votes on the two bills in either house. The lower house
passed the Land Bill, amended, near midnight on March 1,
with one-third of the members absent and little more than
twenty-four hours remaining to the Twenty-second Congress.
(March 3, normally the final day, fell on Sunday.) The Senate
hurriedly concurred, sent the bill to the president, and stayed
in session until the sun rose in order to respond promptly to
the veto many anticipated. But the bill was not returned. Neither was it signed. Jackson resorted to a pocket veto. When
this became known, he was charged in some quarters, northern and southern, with deliberately upsetting the compromise. Clay, although angered by the veto, never took this
view, not even after John M. Clayton made it part of his personal and partisan history of the compromise.[88]

Who voted to extend the olive branch and who the sword?
Usually they were *not* the same persons. Of 188 reresentatives
whose votes were recorded on both tariff and enforcement
bills, 114 voted in opposite ways: 43 for the former but against
the latter, 71 against the former but for the latter. The grand
compromise was passed not by men like Letcher, Clayton,
Grundy, and Rives, who accepted the full responsibility by
supporting both measures, but by men like Tyler and Webster, who did not accept the package but whose converse yea
and nay votes led to its adoption. Paradoxically, compromise
prevailed by mutual antagonism to compromise. It prevailed
though there was no majority for it, only for its separate measures. Clay had counted on this result; Calhoun had acquiesced in it.

88. On the passage of the Land Bill and its alleged relation to the Compromise Tariff, see Clayton to DuPont, March 2, 1833, in Eleutherian Mills Historical Library, and the report of Clayton's speech, June 15, 1844, in Boston
Courier, June 21, 1844. Editorials in the *Courier*, March 9, and the *Mercury*,
March 8, 1833, claim the bill was part of the compromise.

The composite vote of the representatives and senators of the different sections was more or less as expected. The vote of the Northeast (above the Mason-Dixon Line) in the House was 26 to 71. (In the Senate it was 8 to 10.) Only seven congressmen from this section were recorded against the Force Bill. The South (including the Southwest, below the Ohio River) presented much the same division in reverse: 81 to 4 for Clay's bill, 43 to 39 for the Force Bill. The Northwest generally favored both measures, as did Kentucky, Tennessee, and Louisiana. Obviously, Clay failed to convince the congressmen of manufacturing states that his bill would preserve protection. Those of five states (Vermont, Massachusetts, Connecticut, Rhode Island, and New Jersey) voted unanimously against the Compromise Bill; so did all but four of Pennsylvania's twenty-six congressmen. The Clayton amendment secured very few protectionist votes for the bill—it apparently cost no votes on the other side—and although the five-to-one vote of the Clayton mess contributed to the winning margin in the Senate, the bill would surely have passed without it. Obviously, too, neither the Jackson administration nor the Jackson party aroused itself to enact the Compromise Bill. Some of the Van Buren cadre gracefully accepted defeat of the Verplanck bill and voted for Clay's. (The New York vote was 12 to 18.) Party affiliation, however, seems to have had little to do with the division in the Northeast or anywhere else.

The overwhelming southern support of the Compromise Bill reflected the opinion that it was, indeed, a great concession to the South, while the section's strong vote for the Force Bill pointed up the isolation of South Carolina, as well as the continuing popularity of Jackson despite the heresies of the proclamation. The Force Bill, with majority support in all sections, would have been enacted for patriotic reasons if no other, whether or not there was a compromise tariff. That measure, however, could not have been enacted without the Force Bill. If Clay was the masterful strategist of the compromise, he was finally successful only because Jackson wielded the sword behind him.

III Legacy

In the years to come the Compromise Act would generally be celebrated as an act of deliverance, a sacred compact, honorable to its authors, above all to Henry Clay, and a
vindication of American political institutions. It was often the
subject of panegyric. Ohio congressman Tom Corwin, who
had voted for it, remembered the act as nothing short of mi-
raculous. "What, sir, were the happy, the glorious effects of
that compromise?" he asked four years later.

> The day before that law received the President's approval was over-
> cast with the gathering clouds of civil war, deepening, spreading,
> and blackening every hour. The ground on which we stood seemed
> to heave and quake with the first throes of a convulsion, that was
> to rend in fragments the last republic on earth; at this fearful mo-
> ment an overruling Providence revealed the instrument of its will
> in the person of one man, whose virtues would have illustrated
> the brightest annals of recorded time. He produced the great mea-
> sure of concord, and the succeeding morning dawned upon the
> American horizon without a spot; the sun of that day looked
> down, and beheld us a tranquil and united people.[1]

Of course, not everyone shared this glowing opinion of the
act, or of its author. "The principle was bad, the measure was
bad, the consequences were bad," Daniel Webster said after
ten years.[2] And in longer retrospect, Thomas Hart Benton,
the Missouri Democrat, condemned the act as a piece of political jugglery, concocted in secret, managed by desperate political rivals, and carried by rival interests under specious
appeals to national patriotism. "It comprised every title nec-

---

1. *Register of Debates*, 24th Cong., 2nd Sess., p. 1365, in a speech in the
House of Representatives, January 12, 1837.
2. *The Writings and Speeches of Daniel Webster* (Boston, 1903), III, 131.

essary to stamp a vicious and reprehensible act—bad in the matter—foul in the manner—full of abuse—and carried through upon the terrors of some, the interests of others, the political calculations of many, and the dupery of more."[3] The act that was born in the depths of political controversy remained controversial during its legislative life and beyond. Even today historians are confused and divided on some of the most basic questions raised by the compromise.

Of what, in fact, did the Compromise of 1833 consist? Was it embodied in one act or two or, possibly, three? Corwin simply assumed that the compromise and Clay's tariff reform—the Compromise Act as it was promptly named—were one and the same. So did Benton, probably Webster, and with the passage of years nearly everyone else. In truth, however, the Force Act—the sword that attended the olive branch—was necessary to the compromise. This being so, Providence must have chosen not one instrument but two. For surely President Jackson, by his marshaling of federal power to defeat nullification, helped to make the compromise. But the pains and penalties of the Force Bill were never felt, and it was quickly forgotten. The Compromise Act brought relief, and it endured. Moreover, Jackson terrified; he did not conciliate. While he would be remembered for many things, peace and concord were not among them. Clay, who had already won plaudits for his part in the Missouri Compromise, was now celebrated as the Great Compromiser, the Great Pacificator; and he had yet to add still a third jewel to his crown. Whether or not he had saved the protective system—a matter that divided his political friends—even political enemies, like Martin Van Buren, frankly said he had "saved the country."[4] This too, the

3. Thomas Hart Benton, *Thirty Years' View* (New York, 1854), I, 345.
4. *The Autobiography of Martin Van Buren*, in the American Historical Association *Annual Report, 1918* (Washington, D.C., 1920), 560. William W. Freehling, *Prelude to Civil War: The Nullification Controversy in South Carolina* (New York, 1966), 294; and, especially, Major Wilson, "Andrew Jackson, The Great Compromiser," *Tennessee Historical Quarterly*, XXVI (1967), 64–78, go much too far in crediting, and applauding, Jackson for the compromise.

idea that the country had been saved from some catastrophe, while widely shared, was denied by many who had seen no imminent danger from South Carolina and by others who felt that the compromise only postponed an inevitable day of reckoning between North and South. That awful day had been brought nearer by the fatal precedent, in the opinion of John Quincy Adams, of the Union's yielding to nullification.

South Carolina, of course, had its own view of the compromise. John C. Calhoun left Washington on Monday, March 4, traveling day and night over snow-covered and rain-soaked roads, much of the time in open mail-carts, in order to reach Columbia when the state convention reassembled on the eleventh. He felt some anxiety lest the hotheads in the blue-cockade-and-palmetto-button party drive the convention full speed ahead into nullification. There had been reports that several of the Columbia Nullifiers, such as Thomas Cooper, president of the college, who had been the first to call upon South Carolinians to "calculate the value of the Union," and William C. Preston, leader of the party in the legislature, favored this course. Neither had been exposed to the chastening tempers of congressional politics. Calhoun need not have worried. The principal state leaders, including Governor Hayne, James Hamilton, Jr., and Preston, had followed the senator's lead and grabbed at Clay's peace offering. In the convention the state's militant young attorney general, Robert Barnwell Rhett, stood virtually alone in opposing repeal of the nullification ordinance. Expressing his contempt for the Union, as for the compromise, he advocated a southern confederacy. To ease the fears of those who, while agreeable to the Compromise Act, had little faith in Clay's pledge to honor it, Hamilton proposed that if in 1842, or before, protective duties were revived, South Carolina would be free to resist. But the convention, perhaps because it saw no need to declare the obvious, declined the amendment, then repealed the ordinance, 155 to 4.

Before adjourning three days later, the convention passed another ordinance pronouncing the Force Act null and void.

This was more than a face-saving gesture. South Carolina did not recognize the act as part of the compromise; its passage was merely "an ebullition of spleen" or "bravado," in the words of the Charleston *Mercury*, since it could have no operation once the tariff was satisfactorily adjusted. Its objectionable military provisions would expire at the end of the next congress, so it posed no threat. However, because the law stamped CONSOLIDATION on the statute book, because its doctrine struck at nullification, it could not be allowed to stand as a precedent. The tariff, the cause of free trade, was no longer the Palmetto State's primary concern. There was, said George McDuffie, "a yet deeper cause, bringing with it a still more imperious necessity of resistance"—slavery—upon which South Carolina and her sister states must unite. The defense of slavery had always been an element, more or less implicit, in southern resistance to the American System. For the radicals, in fact, the latter was considered "the outer wall" of slavery's defense. Now, with the settlement of the tariff, coinciding with the upsurge of abolitionism in the North, slavery came to the fore. That abolitionism, if it captured northern opinion and connected itself with the consolidating doctrines of the proclamation and the bloody bill, was far more dangerous to the South than the American System had ever been.[5]

The Carolinians not only accepted the Compromise Act as a fair settlement between northern manufacturing and southern plantation interests but also proclaimed it a triumph of nullification. Robert Turnbull, one of the original Nullifiers, was jubilant. "With but our one-gun battery of nullification we have driven the enemy from his moorings, compelled him

5. *Life of John C. Calhoun* (New York, 1843), 48–49; Joel R. Poinsett to Andrew Jackson, February 22 and March 21, 1833, in Papers of Andrew Jackson [Microfilm] Library of Congress; William C. Preston to N. B. Tucker, February 18, 1840, in *William & Mary Quarterly*, 1st Ser., XII (1903), 142–43; Laura A. White, *Robert Barnwell Rhett, Father of Secession* (New York, 1931), 26–27, 75; Richmond *Enquirer*, March 26, 1833; Charleston *Mercury*, March 5, 1833; Pendleton (S.C.) *Messenger*, April 17, 1833; Henry L. Pinckney, *An Oration Before the State Rights and Free Trade Party* (Charleston, 1833); *United States Telegraph*, March 26, 1833.

to slip his cable and put to sea."[6] Officially, the convention re-
sisted the temptation to make this boast, but it was the theme
of countless speeches and editorials and of an elaborate victory ball in Charleston.[7] This celebration in the camp of the
vanquished was aggravating to Clay and his friends. They
had, after all, saved the Nullifiers from Jackson's vengeance
by giving them an apology for retreat, one they had grasped
with the eagerness of drowning men but over which they now
dared to exult in triumph. The South Carolina Unionists, fore-
seeing that this must be the political result in the state, had
been cool to the Compromise Act from the first. It enabled
the Nullifiers to snatch victory from the jaws of defeat. It left
the Unionists, whose patron was Jackson, isolated and vulnerable; and as the passage of time would disclose, it ended
the possibility of healthy two-party competition in South Carolina politics. The heroics of nullification created a legend of
resistance to tyranny in South Carolina that, in association
with the revolutionary one, elevated pride to principle, obsti-
nacy to idealism, and identified the state's defense of slavery
with the pursuit of liberty.

Despite the plain evidence of the defeat of nullification—
a defeat all the more telling because the terms of surrender
forced the state to sanction a national power it had for eight
years insisted was unconstitutional—the claims of victory
could not be begrudged the Nullifiers when so many in the
North acknowledged, lamentably, their validity. Adams, for
instance, thought Clay's act ought to be entitled "An Act for
the protection of John C. Calhoun and his fellow nullifiers,"
for it programmed "the sacrifice of the North to the South, of
the free laborer to the slave holder." The Force Act, on the
other hand, had no effect, exacted no punishment, indeed
was brazenly defied by the convention. "This is the pernicious
and irreparable mischief of Clay's Bill," wrote Adams. "It gave
the show of victory to nullification at the very moment when

6. Quoted in Chauncy S. Boucher, *The Nullification Controversy in South
Carolina* (New York, 1916), 291.

7. *Mercury*, March 23 and 26, 1833; Theodore D. Jervey, *Robert Y. Hayne and
His Times* (New York, 1909), 357–61.

it was at its last gasp."[8] The spectacular rise of cotton prices in the wake of the compromise—60 percent in six months—further emboldened the Nullifiers. It was attributed to tariff reform, though not a single duty had yet been reduced. In November, after a sweeping victory in the congressional election, the State Rights and Free Trade party ceremoniously laid the cornerstone of a monument in Charleston to the deceased Turnbull, then rallied at the Circus in a tribute of loyalty and gratitude to their leader, Calhoun. It was Calhoun, not Clay, not Jackson, who was responsible for the compromise, according to these partisans; and Calhoun vaunted himself on this illusion. Nullification had overthrown the American System, he declared. The guns of Charleston's fortresses were again turned out to sea. But a new crisis had arisen. Under the encouragement of the Force Bill abolition societies had sprouted like mushrooms in the North. The southern states must be aroused to the danger, he declared; and as there could be no union under the bayonet, the repeal of that tyrannical law should be the first order of business of the new Congress.[9] After this political *coup de théâtre* Calhoun's leadership in South Carolina was unassailable. No wonder that he could later say he wanted but one word engraved on his tombstone —Nullification.[10]

Andrew Jackson, the "bloody Richard" of the Nullifiers, believed that the compromise offered only a truce, that it must be succeeded by a new trial of strength. He had never recognized the legitimacy of the South Carolina complaint; personalizing the issue, as was his habit, he had brushed aside the tariff as only a pretext for Calhoun's relentless ambition. Now, sniffing a change in the wind, he concluded that nullification was but a stratagem in a higher game looking to secession and a southern confederacy. "The nullifiers in the South in-

8. John Quincy Adams to Charles Francis Adams, March 26, 1833, in Adams Papers [Microfilm], Massachusetts Historical Society.

9. *Enquirer*, November 26 and 29, 1833; *Niles' Weekly Register*, December 7, 1833; Jervey, *Hayne*, 366–67.

10. Quoted in Sarah M. Maury, *Statesmen of America in 1846* (Philadelphia, 1847), 172.

tend to blow up a storm on the slave question," he wrote. "This ought to be met, for be assured these men would do any act to destroy this union and form a southern confederacy bounded, north, by the Potomac river."[11] It was apparently believed in administration circles that medals had been cast in South Carolina bearing the inscription JOHN C. CALHOUN, FIRST PRESIDENT OF THE SOUTHERN CONFEDERACY.[12]

Jackson had perhaps not altogether dismissed from his mind the idea that had found such varied advocates as Daniel Webster, Joel R. Poinsett (the South Carolina Unionist), and Secretary of State Edward Livingston, of bringing about a political realignment on the grand issue of the preservation of the Union against all selfish combinations, personal or sectional. The combination of Clay and Calhoun, for both of whom Jackson had inexhaustible hatred, was his main concern. This bastard coalition, he suspected, would conspire to recharter the Bank of the United States and by its sinister influence, joined to the corrupt influence of the surplus, doom the country to iniquity.[13] It was of first importance, therefore, to move aggressively against the bank by removal of the government deposits. Although other and more important considerations recommended this course of action, it was in some part a calculated political response to the alliance of Clay and Calhoun consummated in the nullification crisis. Jackson badly miscalculated, however, if he expected to form a broad patriotic front while simultaneously pursuing his war on the bank. As usual, his motives were baffling. Felix Grundy once remarked, "The general is a sportsman and must always have a cock in the pit."[14] Just now it was the bank.

11. Jackson to John Coffee, April 9, 1833, quoted in Marquis James, *The Life of Andrew Jackson* (Indianapolis, 1938), 622. See also Jackson to the Reverend Andrew J. Crawford, May 1, 1833, in John S. Bassett and J. F. Jameson (eds.), *Correspondence of Andrew Jackson* (Washington, D.C., 1926–35), V, 72; and Washington *Globe*, May 1, 1833.

12. *Globe*, May 6, 1833.

13. Jackson to Hugh Lawson White, March 24, 1833, in Bassett and Jameson (eds.), *Correspondence*, V, 46.

14. Quoted by John M. Clayton, in speech at Lancaster, Pennsylvania, September 5, 1844, in *Niles' Weekly Register*, September 14, 1844.

At the same time Jackson sought to reassure states' rights Democrats of his political purity. He vetoed the Land Bill mainly on the ground that creating a fund for distribution to the states would make the states mercenary dependents of the federal government. "A more direct road to consolidation cannot be devised," he declared. "Money is power, and in that Government which pays . . . will all power be concentrated."[15] Some of the president's most eager followers, who admired him for attempting to restore the government to the watergruel regimen deduced from the Jeffersonian "doctrines of '98," were still troubled by the heresies of the proclamation. Since January the editor of the *Globe*, Francis P. Blair, had been introducing glosses on the disputed doctrinal points in an effort to satisfy Thomas Ritchie and other guardians of a warped tradition. They were centered in Virginia, where the legislature debated resolutions denouncing the proclamation and lawyer-politicians like Littleton Waller Tazewell published learned disquisitions in defense of state sovereignty, state interposition, the right of secession—every right claimed by South Carolina except the right of nullification.

While vacationing at the Rip Raps in August, the president read in a Norfolk newspaper that the venerable old Republican, Nathaniel Macon, in retirement in North Carolina after a career of nearly forty years in Congress, had said that the proclamation was as heretical as Calhoun's theory of nullification. Jackson wrote to ask this oracle where he had deviated from the true faith. Macon replied that a sovereign power, such as any one of the United States, could not commit treason or rebellion or be lawfully coerced. The attempt against South Carolina was unprecedented, all the precedents cited by Jackson—suppression of the Whisky Rebellion, enforcement of the embargo—having involved the use of force against citizens, not states. As for Jackson's "perpetual union," Macon said the right of secession had never been surrendered, and

15. J. D. Richardson (ed.), *A Compilation of the Messages and Papers of the Presidents, 1789–1897* (Washington, D.C., 1907), II, 57–68.

he would go to his grave believing South Carolina had that right.[16]

Soon after this exchange there appeared in the *Globe* an "authorized exposition" of the proclamation. While it retracted nothing that was said in the original, it altered the idiom so as to reconcile, superficially, the proclamation with states' rights doctrine. Thus the idea that the Constitution was the creation of "one people," always the keystone of nationalist theories, receded before the Madisonian conception of a constitution formed by the people of the United States acting as separate sovereign communities. Thus the language of "compact," which the proclamation treated with indifference, was returned to favor; and although both nullification and secession were rejected, the right of state interposition on the principles of the Resolutions of '98 (which the proclamation had neglected to mention) was clearly affirmed. Despite the imprimatur, there is no evidence that Jackson actually authorized this exposition. It was an act of folly, he told Van Buren, to explain a proclamation that needed no explanation and that "99 out of every hundred" citizens had instantly approved. Nor could it fairly be said that the exposition "explained away" or "recanted" the objectionable doctrines of the proclamation. Still, Virginia Jeffersonians, like Ritchie, warmly received it as such, and the idea of Jackson's recantation entered into southern apologetics of secession. The "counter-proclamation" was little noticed elsewhere. Not a single newspaper north of the Potomac published it, according to a prominent advocate.[17]

In northern manufacturing states, whose congressmen had voted so overwhelmingly against Clay's bill, the compromise

16. The letters in this correspondence, dated August 17 and 26, September 2 and 25, are all in Jackson Papers.

17. *Globe*, September 27, 1833; Jackson to Van Buren, September 29, 1833, in Bassett and Jameson (eds.), *Correspondence*, V, 212. See Condy Raguet's remarks in his *Examiner and Journal of Political Economy* (Philadelphia), October 16, 1833. I have discussed aspects of the "authorized exposition" in *The Jefferson Image in the American Mind* (New York, 1960), 60–61. Edward A. Pollard, *The First Year of the War* (2nd ed.; Richmond, 1862), 14, considered the exposition significant. It was reprinted in Alexander H. Stephens, *Constitutional*

was at first perceived as a betrayal of the American System, of the section's economic interests, and of the Constitution. In Philadelphia Mathew Carey, since 1819 the foremost publicist of protectionism, penned his "valedictory" and went into retirement. "I am sick, sick, sick, of the prospects of the Union," Carey wrote.[18] In Boston Joseph T. Buckingham, who had launched the Boston *Courier* in 1824 as the premier New England organ of Henry Clay and the American System, denounced the compromise and switched his allegiance to Daniel Webster. Buckingham critically examined the Compromise Act, both its economic and constitutional aspects, in a series of articles entitled "Mr. Clay and the American System." Without impeaching Clay's motives, he expressed amazement at the revolution in the senator's opinion and insisted that the act abandoned protectionism. Death was no less certain for being gradual. In 1842 uniform *ad valorem* duties, without protection or discrimination, would prevail; this had always been the free traders' platform. The big manufacturers, especially in cotton, would survive, but the mass of small ones would fall before foreign competition, Buckingham predicted. The woolens industry would be destroyed. He doubted the advantage or practicality of home valuation. The "miserable juggling" that produced the act, the ambiguity surrounding the question of a pledge showed an opportunism and carelessness toward the future strikingly at odds with the high principles and enlarged vision that had distinguished Clay's statesmanship. Distressingly, this revolution of policy occurred before the empty threat of a single state, for nullification would have passed "as harmless as the lightning of the firefly" without the compromise. Bad as the bill was with respect to the principles of political economy, Buckingham contin-

---

View of the Late War Between the States (Philadelphia, 1868–70), I, 462–72. See also Henry S. Foote's remarks in the Senate, *Congressional Globe*, 31st Cong., 1st Sess., 1496.

18. Mathew Carey to Josiah Johnston, March 21, 1833, in Josiah S. Johnston Papers, Historical Society of Pennsylvania. See also Washington *National Intelligencer*, March 16, 1833.

ued, "its worst features are the virtual surrender which it makes of the principles of the Constitution." The Force Act was of no account since the principle of the law it was meant to enforce had been abandoned. "Talk of the danger of a consolidated government!" Buckingham exclaimed. "The government practically established by this bill has no solidity; it has neither bones, sinews or muscles; it is the mere skeleton of a shadow."[19]

The analysis gave no credence to the idea that the manufacturing interest was one of the parties to the Compromise of 1833. This idea was widely held in the South; it was repeatedly voiced by Calhoun; and it was easily inferred from the fact that the negotiators of the compromise were spokesmen of great sectional interests, plantation agriculture and industrial manufacturing. But, of course, whatever might be said as to the former, the latter had not been a party to the compromise. "Who were the parties to it?" Nathan Appleton asked rhetorically. "The Jackson party and the nullifiers of South Carolina. Mr. Clay acted the part of mediator between these hostile parties who united in carrying it through notwithstanding the opposition and remonstrances of the manufacturing interest." As to that interest, "They were the victims, the sacrifice offered up on the altar, to consummate the reconciliation of two hostile parties."[20] Conceding the truth of this, it would nevertheless be very difficult to show that the Jacksonians were one of the "hostile parties" to the compromise. Benton, who voted against the measure, probably voiced the majority opinion of his party in the denunciation he wrote twenty years later; yet he agreed with Appleton that one of the adversaries, "the manufacturing interest," had not consented or, indeed, even been consulted. "To call this a

19. Boston *Courier*, April 24–27, 29–30, March 4, 1833. See also Joseph T. Buckingham, *Personal Memoirs and Recollections of Editorial Life* (Boston, 1852), II, 1–11 and *passim*.

20. Nathan Appleton to Abbott Lawrence, February 15, 1841, in Nathan Appleton Papers, Massachusetts Historical Society; speech in the House of Representatives, July 5, 1842, reported in the Boston *Courier*, July 9, 1842.

'compromise' was to make sport of language. . . . It was like calling the rape of the Romans on the Sabine women, a marriage."[21]

Here too, in these reflections, the compromise had been reduced to the Compromise Act. Although many of the protectionist critics were never reconciled, the act rapidly gained acceptance in northern opinion. This was especially true in New England. (Pennsylvania was less forgiving.) Clay's friends, returning from Congress, corrected the widespread impression that leading spokesmen of New England interests, like Webster and Appleton, had not been consulted. They also availed themselves of the sudden rise in manufacturing stocks to promote the compromise. The gloomy apprehensions of men like Buckingham disappeared on the tide of prosperity. In the spring Abbott Lawrence, the Boston capitalist, wrote to Clay that earlier objections had been overcome and the manufacturers now believed the act both wise and fair.[22] This was an astonishing reversal, which left many congressmen embarrassed to account for their votes. As Clay had argued and as manufacturers began to recognize, the Compromise Act afforded the enormous advantage of a secure and stable tariff removed from politics. And the reduction of duties, once they began, would work so gradually as scarcely to be noticed in a booming economy. The law became popular even in protectionist quarters.

The compromise endured, in part, because no political Hercules lifted his club against it. Webster was the obvious candidate for the role. Viewing the compromise as "an attempt to make a new Constitution," he never held himself bound by it.[23] Returning home in March he took along the notes for his last, and undelivered, speech against the bill with the intention of preparing them for publication. Stopping at Philadel-

21. Benton, *Thirty Years' View*, I, 312.
22. Peleg Sprague to Clay, March 19, 1833, in Calvin Colton (ed.), *The Works of Henry Clay* (New York, 1904), V, 354–56; Abbott Lawrence to Clay, March 26, 1833, *ibid.*, 357–58.
23. Notes for Speech on the Sub-Treasury Bill, March 12, 1838, in Papers of Daniel Webster [Microfilm], Dartmouth College.

phia, however, he was dissuaded by a friend who urged the importance of harmony between the foremost opposition leaders. The friend was almost certainly Nicholas Biddle, president of the Bank of the United States, whose intercession had been requested by Clay. "I wish you would say all you can to soothe him," he wrote. "You hold a large flask of oil and know how to pour it out." Biddle, even more anxious than Clay to prevent an estrangement, artfully applied the soothing oil, kept Webster from any public platform, and reported that while the senator had not changed his opinions, he had departed in an amicable frame of mind.[24] Clay himself, though he supposed that Webster indulged the hope he would immolate himself on the altar of compromise, was satisfied there was no breach between them. In April an article with an authoritative air turned up in the press denying the rumored breach as well as the persistent tale that Webster had gone over to the administration.[25] But the rumors would not die down.

In June Webster commenced a heralded western tour with a view to broadening his political constituency. Some observers suspected he had a more precise purpose: to render the compromise odious in Clay's backyard. Everywhere he was hailed as the Defender of the Union and the Constitution. Jackson's friends, far more than Clay's, it was often said, paid him court; and, as it happened, Webster's friends were paying court to the touring president in New England at the same time. The senator's speeches at Buffalo and Pittsburgh were interpreted in some quarters as manifestos for the restoration of the American System.[26] But if Webster ever intended to challenge the compromise, he failed to follow through. He was in a state of indecision about his political course. When

24. Webster to Hiram Ketchum, January 20, 1838, *ibid*; Clay to Nicholas Biddle, March 4, 1833, in Typescripts, Papers of Henry Clay, University of Kentucky; Biddle to Clay, March 25, 1833, in Colton (ed.), *Works*, V, 356–57; Clay to Biddle, April 10, 1833, in Typescripts, Clay Papers.

25. Copied in the Boston *Courier*, May 15, 1833, and *National Intelligencer*, May 21, 1833.

26. *Examiner and Journal of Political Economy*, August 7, November 13 and 27, 1833.

he returned to the East he resumed negotiations that, if successful, would have put him in the Democratic party or, he
hoped, in some grand party of the Union. These broke up the
rock of financial policy, however. When Jackson proceeded in
the new Congress to destroy the country's banking and currency system, Webster resumed his place in the opposition. [27]

The cholera epidemic in the Ohio Valley cut short Webster's
tour before he could meet Clay in Lexington. In the fall, after the epidemic subsided, the Kentuckian made a circuit of
the eastern cities, his first in fifteen years. At Boston, where
he pled eloquently for the compromise, he was reportedly
snubbed by Webster. If so, it made no dent in his armor; and
he returned to Congress satisfied that he had lost little of his
following in the eastern states. [28] Whether he had gained any
in the South remained to be seen. Politics did not run on gratitude, and having nothing more to offer the South, unless he
chose to become a zealot for slavery, Clay was skeptical of his
prospects in that section. Indeed, the political rewards for the
statesman who had "saved the country" were poor everywhere, just as they had been after the Missouri Compromise.
Some of his oldest and closest associates turned elsewhere for
a presidential candidate to oppose Jackson's handpicked successor, Van Buren. [29] Clay kept this option open but was not
hopeful. His purpose when Congress convened was to unite
the fragments of the opposition in a war against "Jackso-
nism," a disease more fatal than the cholera, in his opinion.
Jackson's high-handed action in removing the federal funds

27. See Webster's "Memorandum on the Nullification Crisis, 1830–33,"
[1838?], in Webster Papers, and the account in Norman D. Brown, *Daniel
Webster and the Politics of Availability* (Athens, Ga., 1969), Chap. 3.

28. *Examiner and Journal of Political Economy*, October 30, 1833; Joseph S.
Jones to David Swain, October 26, 1833, in David Swain Papers, University
of North Carolina; Robert C. Winthrop, *Memoir of Henry Clay* (Cambridge,
1880), 30–31.

29. For instance, John M. Clayton to John McLean, October 4, 1833, in
John McLean Papers [Microfilm], Library of Congress; John J. Crittenden to
L. L. Nicholas, September 13, 1833, in John J. Crittenden Papers [Microfilm],
LC; Elisha Whittlesley to Daniel Webster, September 14, 1833, in Webster
Papers.

from the Bank of the United States, sending tremors through the business community and further alienating members of his own party, offered another opportunity, much more promising than the tariff, to draw together the antipodes, "the nationals" and "the state rightists," into a strong new party.

The Whig party was the result. It was badly divided, not only from its schizophrenic origins but also because of the rivalry among its great men. The coalition between Clay and Calhoun, having begot the party, became symptomatic of its problem. It was, as everyone recognized, "a bundle of opposites" held together only by hostility to Jackson. Even in their common assault on executive tyranny, the leaders proceeded from opposite premises. For Clay the usurpation lay in the destruction of a system of national legislation and in the aggrandizement of executive power at the expense of congressional and judicial power. It was this that provided the symbolic link to parliamentary opponents of Stuart kings and justified the Whig name. For Calhoun, on the other hand, executive usurpation had its source in congressional usurpation, in the consolidation of national power, to which states' rights offered the only effectual check.[30]

The coalition was thus an embarrassment to both of the principals, so much so that victory over the Jacksonians loomed as a disaster to one or the other, perhaps to both. The third member of "the great triumvirate," Webster, was still unsure of his course when the Twenty-third Congress convened. Pride and temper and ambition, far more than principle or policy, brought him into conflict with Clay. At every opportunity, it was observed in Washington, "they shiver a lance with each other." Webster's bargaining with the administration had stirred Clay's distrust, while he, in turn, could never forgive Clay for his alliance with Calhoun. Even as Webster fell into line, again submitting to Clay's leadership in

30. For a typical statement of Clay's view, see his Speech on the Expunging Resolution, January 16, 1837, in Daniel Mallory (ed.), *The Speeches of Henry Clay* (New York, 1843), II, 264–78. Especially revealing of Calhoun's view is his public letter to Thomas Walker Gilmer in *Niles' Weekly Register*, August 9, 1834.

Congress, he struck out for the presidency on his own, with no help from either of the fellow triumvirs.[31]

In the "panic session" the issues of tariff and surplus, of nullification and compromise, were almost forgotten. Faithful to his promise, Calhoun proposed the repeal of the Force Act. It furnished an occasion for philosophical discourse, of course. Calhoun's speech was especially interesting for its view of the latest act of executive usurpation, the removal of the deposits, as only an extension of the former one. Clay, as if to make up for his silence when the bill was enacted, spoke at length against repeal. By asserting that such action would give grounds for the boasted triumph of nullification, he seemed to acknowledge that the Force Bill had been an integral part of the compromise. This was not his main plea, however. "Sir," he said, "we have got a nullification infinitely more dangerous, not in South Carolina, but in Washington, which threatens destruction to the liberties of the country, by an entire absorption of all the power of the General Government in the hands of one man. The nullification of our Southern sister was bygone. Is this, then," he asked, "the time for abstract propositions and metaphysical discussions? It would be better to let them sleep."[32] The repeal bill was sent to the Judiciary Committee for burial. Earlier, in the House of Representatives, a motion instructing the Ways and Means Committee to report a plan that would repeal the nine years of protection provided by the Compromise Act was rejected on a vote of 69 to 115.

The Compromise Act again came into discussion when Congress faced the dizzily spiraling surplus in 1835 and 1836. Customs revenue continued to rise, at the rate of three million dollars a year; and proceeds from the sale of public lands rocketed from four million in 1833 to almost twenty-five million three years later. The administration directed the deposit

31. Boston *Courier*, December 20, 1833; Willie Mangum to David Swain, December 22, 1833, in Swain Papers; Brown, *Webster*, Chap. 4.

32. *Register of Debates*, 23rd Cong., 1st Sess., Senate, p. 1281. Calhoun's speech is in Richard K. Crallé (ed.), *Works of John C. Calhoun* (New York, 1853–55), II, 376–405.

of all this money in specially designated state banks, which were chosen as much for their political as for their financial repute and were, therefore, called "pet banks" by the opposition. In December, 1833, there were thirty-three of them holding a Treasury surplus of twenty million dollars. And the surplus was growing by approximately that amount annually. In 1835 it actually exceeded the government's expenditures. The potential of this money power for injury to the economy was well recognized. So, too, was the danger of political corruption and influence sapping the foundations of republican government. What was to be done with the surplus?

One obvious solution to the problem was drastic reduction of the tariff; but this would violate the compromise, already bathed in an aura of sanctity, and it found no forthright advocate in Congress. Clay continued to push his Land Bill. Along with recharter of the bank and adherence to the compromise, it was a crucial measure in his political platform. Under his bill the accumulated surplus, pegged at twenty-one million dollars, would be immediately distributed to the states; and all but 10 percent of the proceeds from land sales for two more years. The latter would have the same effect as a distribution of the surplus, of course. But for Clay the Land Bill had two larger purposes: first, the securing of the public domain, this national trust, from the ravages of speculators, squatters, and politicians; second, the prosecution under state auspices of works of internal improvement the Jackson administration had rejected under the federal head. Still angry at the president for his pocket veto of the bill passed in 1833, Clay spoke lyrically of the good it would have done. "What immense benefits might not have been diffused throughout the land! . . . What new channels of commerce and communication might not have been opened! What industry stimulated, what labor rewarded!"[33] The Jackson administration and the Democratic party opposed Clay's solution, though they had no compelling alternative. Ceding or giving away the public lands was popular only in the western states. Benton provided the con-

33. *Register of Debates*, 24th Cong., 1st Sess., Senate, p. 51.

gressional leadership for an administration plan to pour millions of dollars into fortifications and other works of national defense. A foreign crisis with France, stemming from a quarrel over the payment of old spoliation claims, raised the credibility of this policy in the new Congress; but it offered only marginal relief at best and disappeared when the crisis was amicably resolved.

Another plan, which Congress finally adopted, belonged to Calhoun. His fears of consolidation and the rule of a despotic sectional majority in Washington turned on a theory of the dominance of economic interests in politics. The protective tariff exemplified the theory. But he also held, somewhat inconsistently, that the majority was viciously manipulated by partisan demagogues and spoilsmen, which made democratic power as corrupt as it was self-serving.[34] This was shown strikingly, Calhoun believed, in the numerous abuses of the patronage power under Jackson. The government had been turned into a patronage machine subservient to a political party, a party held together not by principle or policy but by hope of reward, extending over the whole country the system of spoils politics Van Buren and associates had first matured in New York. In his influential Report on the Extent of Executive Patronage, in February, 1835, Calhoun had argued that the only permanent remedy for these newer political evils was the same as for the older ones: radical retrenchment of the revenue. But this was forbidden immediately by the Compromise Act. Therefore, he proposed, let the Constitution be amended to permit distribution of the surplus revenue to the states until 1843, when the goal of a revenue tariff would be reached.[35] The proposal came as a surprise, for Calhoun had opposed the distribution plan as inherently centralizing when Jackson recommended it. But true consistency, as he once re-

34. See William W. Freehling's perceptive article, "Spoilsmen and Interests in the Thought and Career of John C. Calhoun," *Journal of American History*, LII (1965), 25–42.
35. Crallé (ed.), *Works*, V, 140–90.

marked, showing his capacity for opportunism, "is to act in conformity with circumstances."[36] In this case he could point to the mountainous surplus, the restraints imposed by the Compromise Act, and, as he saw it, the rapid degeneration of the government. An additional circumstance, though rarely acknowledged, was South Carolina's interest in the surplus to finance a projected western railroad.

Nothing came of this plan, nor was there any hope for it as long as it took the form of a constitutional amendment. In April, 1836, after months of inconclusive debate in Congress on Clay's Land Bill, on fortifications, and so on, Calhoun abandoned the plan but proposed to achieve the same result by an ingenious provision inserted into his bill for regulation of the federal deposits in the pet banks. Under this provision the surplus in excess of five million dollars would be "deposited" in the state treasuries in proportion to the federal population ratio. Calhoun heatedly denied that this was distribution of the surplus under another name. Although the states would enjoy free use of the deposits, they could be recalled in a national emergency, and so had the character of interest-free loans. To this Benton snapped, "Names cannot alter things; and it is as idle to call a gift a deposit, as it would to call a stab of a dagger a kiss of the lips."[37] Nevertheless, the Deposit Bill passed through Congress and received Jackson's signature under this fiction.

Clay, seeing that his own bill was bottled up by the administration majority in the House after again clearing the Senate, voted for the deposit bill because it embodied substantially the same principle and purpose. Calhoun extolled the act as "the commencement of a new political era . . . marking the termination of that long vibration of our system towards consolidation." It returned the government to its original confederate principles. It was "the consummation of the Carolina

36. *Ibid.*, IV, 268.
37. *Register of Debates*, 24th Cong., 1st Sess., Senate, p. 1810. For Calhoun's speech, see pp. 1616–35.

doctrines," since it placed the nation's Treasury in the keeping of the state governments. [38] Benton, his archenemy, agreed the act was revolutionary. But whether the revolution was in the direction of states' rights or of consolidation, as Calhoun himself had earlier feared, was problematical. [39]

In his last annual message to Congress, Jackson criticized the workings of the Deposit Act and called for immediate reduction of the revenue to the wants of the government, thus openly inviting attack on the Compromise Act. Compromise-busting measures were introduced in both houses by political intimates of the president-elect, Martin Van Buren. Clay and Calhoun—Webster, too—stood resolutely against any violation of the compromise. The former, with obvious pride, said the act was one of three great compromises in the nation's history; like the Constitution and the Missouri Compromise, it had saved the Union and taken the character of a sacred compact. No statute could bind Congress, of course; but neither would the compromise have been possible without the plighted public faith to maintain it. The only tariff tampering allowed by the act was with duties below 20 percent on non-protected articles. Because the bills in the Senate and House went well beyond this, they must be rejected.

Calhoun, no less forthright in his defense of the compromise, saw in these maneuvers a devious political game to re-embroil North and South on the tariff. He recalled the infamous Tariff of Abominations of 1828, in which Van Buren,

38. Letter to the Citizens of Athens, Georgia, August 5, 1836, in *Niles' Weekly Register*, August 27, 1836; Speech on the Deposit Act, *ibid*, October 1, 1836. For Clay's view, see his speech at Woodford, Kentucky, *ibid.*, September 3, 1836.

39. The Treasury "deposited" with the states twenty-eight million dollars in three quarterly installments, from January through July, 1837. Because of the financial crisis that commenced in the spring, the fourth installment was deferred indefinitely and the bill was not renewed by Congress. The deposits were never withdrawn. In view of the size of the surplus, and the amounts distributed in comparison to either federal or state budgets of the time, the experiment was potentially much more revolutionary in its effects than the federal "revenue sharing" of recent years. A new history of the surplus and distribution is needed. Meanwhile, see Edward G. Bourne, *The History of the Surplus Revenue of 1837* (New York, 1885).

Silas Wright (sponsor of the present Senate bill), and other northern Jacksonians had tricked southern congressmen into voting for outrageously high duties in the belief they would cause the bill to be defeated by New England votes. But the bill had passed, as intended. Now the South was offered small favors at northern expense. Why? In order to reopen the entire system, Calhoun answered, to revive the politics of "tariff juggling," and to renew exploitation of the South. "Let me tell my Southern friends," he declared with his piercing eyes on the Jackson leaders, "that I know the men with whom we have to deal. Abandon the compromise, and they will be among the first to resist all future reduction. We see the bait, but we do not perceive the hook that lies under it. . . . But I can tell these fishers, they shall not catch me. I was caught once following the same lead. I am not to be caught a second time."[40] The bill in the House was the more alarming one. It drew from the Massachusetts legislature resolutions hymning the praise of the Compromise Act, which the state's entire congressional delegation, including the senators, had voted against in 1833.[41] Neither of the bills became law. As a result of the debate the compromise had strengthened its hold on the political affections of the country.

That debate was the last hurrah of the coalition. The economic boom ended in the Panic of 1837. When the banks, the pets included, suspended specie payments on their notes the government was placed in the predicament of having no safe or lawful depositories for its money. Van Buren called a special session of Congress to deal with the crisis. He recommended that the government conduct its business in specie and establish depositories for its funds independently of the banks. This plan for an Independent Treasury at once struck a responsive chord in Calhoun. Although he had never been identified politically with hard money or the Jacksonian attack on banks, he now saw in the separation of the govern-

40. *Register of Debates*, 24th Cong., 2nd Sess., Senate, pp. 975–76.
41. Boston *Daily Advertiser*, February 2, 1837; Boston *Courier*, February 18, 1833.

ment from banks the triumphal arch opening into the promised land of states' rights under the Jeffersonian flag of '98. Banks fed on consolidation; consolidation fed the money power of banks. Dissolve the connection and the government would be restored to its original principles more surely than by the destruction of the protective tariff. The Whigs, with whom Calhoun had acted for four years, opposed the administration plan. Most of them backed Clay in his call for a new national bank. When William C. Preston, South Carolina's junior senator, suggested that Calhoun had left the Whigs to become "an administration man," he coolly replied that he was never a Whig, though he had cooperated with them in the struggle against executive tyranny. Nor was he a Democrat. "I belong to the smallest party in the country; I am simply an honest Nullifier." But in this crisis the new Democratic administration, with no place else to go, had fallen back on the old Republican platform, and he, of course, availed himself of this extraordinary opportunity to accomplish his object. He had not gone over to the administration; the administration had come round to him.[42]

After Congress adjourned in the fall, Calhoun explained himself to his South Carolina constituents through a public letter in the Edgefield *Advertiser*. He recalled the crisis of 1833, when nullification had overthrown the American System, and said he had then joined his former enemies from necessity to put down executive tyranny. But the Panic of 1837, coincident with Van Buren's accession, had completely altered the political landscape. The administration was thrown on the defensive. "It was clear that, with our joint forces [Whigs and Nullifiers], we could utterly overthrow and demolish them," Calhoun wrote. "But it was not less clear that the victory would enure not to us, but exclusively to our allies and their cause." By supporting the administration's effort to divorce bank and state, on the other hand, political parties would be forced

42. Crallé (ed.), *Works*, III, 60–61; *Register of Debates*, 25th Cong., 1st Sess. (Special), pp. 276–77.

back on "the old and natural division of state rights and national," which was most advantageous for the South.[43]

Calhoun's political summersault, whatever his intention, was scarcely less astonishing to his states' rights friends than to "the nationals" who controlled the Whig party. Until the Edgefield letter, it was said, the Independent Treasury had not ten advocates in South Carolina; all at once it had many, and the conversion Calhoun worked in the legislature at Columbia on his return to Washington in December was compared to that of the apostle on the road to Damascus.[44] Presumably the converts were persuaded by the logic of Calhoun's argument that there was greater security with the Democrats than with the Whigs. Banking was only part of it. The same principles that supported a national bank could also support a return to protectionism. Abolitionism was a more potent force in the Whig party; and Calhoun was distressed by the opposition of leaders like Webster and Clay to his efforts to suppress abolitionist agitation. But men in South Carolina, as elsewhere, inevitably wondered if Calhoun's demarche on the Independent Treasury was not motivated by his old passion for the presidency, which might be more readily gratified in succession to Van Buren than in succession to any of the eminent Whig statesmen.

The "day of settlement" between Clay and Calhoun—one of the memorable parliamentary encounters in American history—occurred in February, 1838. Clay, impetuous by nature, had restrained himself during the special session, expecting Calhoun would quickly return to the fold when he discovered that no one was following; but the Edgefield letter was a declaration of apostasy from the Whigs and, backed by the rally at Columbia, it angered Clay. Suddenly, in the midst of a speech against the Independent Treasury, he turned from reasoned argument to personal attack on his former ally. The ar-

43. *Niles' Weekly Register*, December 2, 1837. (The letter was first published on November 3, 1837).

44. James Hamilton, Jr., as reported in *Niles' Weekly Register*, June 30, 1838.

duous campaign they had waged together was about to end
in victory, he said, when Calhoun went "horse, foot, and dra-
goon" to the enemy. There had been nothing like it since
Achilles abruptly withdrew his army from the Siege of Troy.
But Achilles had been wronged; the Whigs had done no wrong
to Calhoun, rather had given scope to his genius and relied
on his fidelity. Still he left. "He left us, as he tells us in the
Edgefield letter, because the victory which our common arms
were about to achieve, was not to enure to him." This, said
Clay, all injured innocence, was the first time he had heard
that personal and party gain was the basis of their alliance, or
a just cause for ending it.[45]

Calhoun took his time to reply. His carefully composed
speech "smelt of the lamp," Benton said, and reminded him
of Demosthenes' Oration on the Crown. He defended his political character and, having first censured Clay for argument
ad hominem, returned the shot with twice the force. As to the
imputation of base motives, founded on the Edgefield letter,
Calhoun insisted his only purpose was to save the cause of
states' rights from being swallowed up. "I stamp it, with scorn,
in the dust. I pick up the shaft, which fell harmless at my feet.
I hurl it back. What the Senator charges on me unjustly," he
declared with his eye fixed on Clay at the opposite side of the
chamber, "*he has actually done*. He went over on a memorable
occasion, and did not leave it to time to disclose his motive."[46]
Calhoun thus revived from several years' sleep the old canard
of "corrupt bargain" between Clay and John Quincy Adams
in 1825.

Replying at once, the Kentucky senator reflected disparag-
ingly on the quick turns and reversals that had marked Calhoun's political course, and marveled at the delusion that now
led him back into the arms of Van Buren and the Democrats.

---

45. Mallory (ed.), *Speeches*, II, 310–49. Two standard accounts are Benton,
Thirty Years' View, II, 97–112, and Nathan Sargent, *Public Men and Events* (Philadelphia, 1875), II, 29–50. See also O. H. Smith, *Early Indiana Trials, and
Speeches, Reminiscences* (Cincinnati, 1858), 245–50, and [Theodore N. Parma-
lee], "Recollections of an Old Stager," *Harpers Magazine*, XLVII (1873), 758–60.
46. Crallé (ed.), *Works*, III, 244–79.

He poured ridicule on the assertion in the Edgefield letter that nullification had destroyed the American System. Far from being extorted by nullification, the Compromise Act had been a compassionate concession to the weakness and folly of nullification, Clay said. Touching on the history of the compromise, he recalled that Senator Clayton had come to him pleading for Calhoun and his beleagured band: "They are clever fellows, and it will never do to let old Jackson hang them." It was in this spirit, Clay asserted, that he had offered the compromise. The idea that it abandoned protection was another of Calhoun's delusions. He had, in fact, been compelled to sanction it constitutionally, and at every step in the making of the compromise, from the term of the law to home valuation, the Nullifier had yielded. A rejoinder by Calhoun, then another reply and rejoinder concluded the affair.[47] Now Webster drew his sword. In a remarkable speech punctuated by droll wit and ridicule, he cut up and scattered Calhoun's vaunted consistency.[48] Their feelings toward each other had always been friendly, though on the point of constitutional doctrine they were more inflexibly opposed than Clay and Calhoun. They talked across a void. Against both Whig giants, Calhoun stood up very well; but the character of a trimmer and apostate was scarcely an enviable one, and he could not escape it.

A dramatic replay of "the conflict of giants" occurred on January 3, 1840. Three weeks earlier, at the opening of the new Congress, Calhoun's followers in the House (the South Carolina representatives and several others) had voted with the administration in a crucial party contest over the New Jersey delegation. Just before the New Year the senator had arranged through a mutual friend to be taken to the White House and presented to Van Buren, a man with whom he had not spoken for almost a decade and who he had bitterly denounced as a Judas, a Janus, and a Catiline.[49] Although he

47. The rejoinders are in Benton, *Thirty Years' View*, II, 112–18.
48. *Writings and Speeches of Daniel Webster*, VIII, 162–237.
49. *National Intelligencer*, September 19, 1840.

had changed his tone during the last two years and worked closely with the administration in financial matters, he had hewed an independent course in other areas and kept a personal distance from the president. If the embrace at the White House was the consummation of the demarche begun by Calhoun at the special session, the secret articles of the new alliance were perplexing. There had been nothing like it since the infamous coalition of Lord North and Charles James Fox in 1783. And Clay might hope that the public outrage, which in that instance had opened the way to William Pitt, would work in the same fashion for him.

Addressing himself to a bill introduced by Calhoun to cede the public lands to the states in which they lay, Clay wondered if it could be part of the recent accord. Inasmuch as the senator had "made his bow in court, kissed the hand of the monarch . . . and agreed henceforth to support his edicts," the public had a right to know what pledges and compromises had been entered into. Calhoun became indignant at the suggestion his personal relations had anything to do with his political principles and policies. He had made no pledges or compromises—Clay was the expert on that subject! After chiding Calhoun for once again resurrecting a political libel that had mercifully sunk into oblivion, Clay replayed his little speech on the virtues of compromise and said that but for one particular compromise the senator from South Carolina would not now be orating in the Capitol. Calhoun jumped to his feet to repel the insinuation that he, or South Carolina, or the Union was under a debt of gratitude to Clay for the Compromise Act. "The senator was then compelled to compromise to save himself. Events had placed him flat on his back, and he had no way to recover himself but by the compromise." He recalled how Jackson, by his proclamation and subsequent message, had rallied the leading friends of Clay's system, including his great rival from Massachusetts, who would have reaped the honors had the contest come to blows. "Compromise was the only means of extrication. He was thus forced by the action of the State, which I in part represent, against his system, by my counsel to compromise, in order to save himself. I had the mastery over him on the occasion." Clay, who had come to believe Calhoun would "die a traitor or a madman," was astounded by this fanciful history. "The senator says, I was flat on my back, and that he was my master. Sir, I would not own him as my slave." A roar went through the gallery. "He my master! and I compelled by him! . . . Why, sir, I gloried in my strength." And he went on to say, as he had said from the beginning, that he had introduced the compromise to save the Union, save the American System, and restore peace and harmony to a divided country.[50]

Washington observers thought they saw in this confrontation foreshadowings of the campaign for the presidential sucession, not in 1840, but in 1844. Several weeks earlier the Whigs had nominated William Henry Harrison to challenge Van Buren in his troubled bid for a second term. In doing so they had turned their backs on Clay, though it was understood that the sixty-nine-year-old general would serve but one term and that Clay would probably be the heir apparent, as Van Buren had been for Jackson. No one doubted that Calhoun would seek the Democratic nomination, whether or not pledges had been given in the December accord. In 1840 he steered South Carolina into the Van Buren column—only one of seven states to vote for him—and this strengthened his claim on the nomination in 1844. Everything thus seemed to be building to a climactic battle of the giants. The third member of the once-powerful triumvirate, Webster, had failed a second time in his quest for the Whig nomination. Much admired for his talents, he could not generate popular enthusiasm. The seeds of distrust sowed in 1833 continued to disturb his relationship with Clay. He chafed under the Kentuckian's commanding, often arrogant, leadership in Congress; and behind the facade of harmony neither man felt very generous toward the other. Webster continued to fret over the Compromise Act, not only because it had "made a new constitution,"

50. *Congressional Globe*, 26th Cong., 1st Sess., 96–98; Benton, *Thirty Years' View*, 119–23; Clay to Harrison Gray Otis, June 26, 1838, in Typescripts, Clay Papers. See also the reports in the Boston *Courier*, January 8, 1840, and the Louisville *Daily Journal*, January 13, 1840.

one more Jacksonian than Whig, but because its history had become the sport of politics. According to Clay, his heroics alone had saved the Union and, amazingly, protectionism at the same time. This turned the bargain entirely to his own account. It was, Webster thought, all humbug and nonsense, though he didn't dare say so. But in January, 1840, having worried over it for two years, Webster authorized the publication of the notes for his final, undelivered speech against the compromise bill.[51]

Because of the panic and the ensuing depression, which rapidly turned Treasury surpluses into deficits, the ambiguities of the Compromise Act assumed tangible importance. The revenue standard, it now appeared, might very well require a rate of duties above the targeted 20 percent. Was this permissible under the compromise? Was protectionism to be revived at the hour appointed for its death? What were the implications of this for Clay's distribution (land) bill, which Whigs increasingly viewed as a conditional part of the compromise? What would happen after June 30, 1842? Would the formerly protective duties continue to be collected at the 20 percent rate, and on the terms of home valuation and cash payment, or would all the duties expire in the absence of new legislation? On this last question, Clay held the former view, one he had clearly stated in his original speech on the compromise bill, though it was left to inference in the statute. Calhoun held the opposite opinion that the act prospectively repealed all duties after June 30, 1842. For several months he had been pressing the Van Buren administration to turn its attention to a new tariff law on the principle of uniform *ad valorem* duties without protection or discrimination. The administration, still struggling for the Independent Treasury, preferred to avoid this hornets' nest.[52] Eagerness for legis-

51. Webster to Hiram Ketchum, January 20, 1840, Stephen White to Webster, January 26, 1840, in Webster Papers; New York *American*, January 22, 1840.

52. Calhoun to R. M. T. Hunter, June, 1839, in Martha T. Hunter, *A Memoir of Robert M. T. Hunter* (Washington, D.C., 1903), 71–72; Calhoun to Joel R. Poinsett, October 16, 1840, in Papers of Martin Van Buren [Microfilm], Library

lation on the southern side was soon matched by rising eagerness in the North. As the depression deepened, industrial interests began to blame the Compromise Act for their difficulties. Pennsylvania was the center of renewed protectionist agitation.[53] In 1844 it would exact terrible revenge on Henry Clay.

The liquidation of the Compromise Act fell to the new Whig administration elected in the celebrated Log Cabin and Hard Cider Campaign of 1840. John Tyler, who succeeded to the executive office upon Harrison's untimely death, boasted he had presided at the birth of the compromise. Of course, he considered it "a solemn compact." In his message to the special session of Congress convened at Clay's behest to enact the Whig financial program, this Virginia Whig of the states' rights school expressed the hope that the Compromise Act, only thirteen months from full term, would become a permanent settlement. This meant, on his understanding of the law, incidental protection under the 20 percent ceiling, and nothing more, unless the exigencies of the revenue required higher duties. He also advocated distribution of the proceeds from the public lands, as long as the ceiling of the Compromise Act was adhered to.[54] The session featured a prodigious struggle between Tyler and Clay over a national bank or some equivalent fiscal agency to replace the Independent Treasury finally approved by Congress only a few months before the Whig landslide buried the Van Buren administration. But Clay, while waging this fight, was also determined to pass a distribution bill, a bankruptcy bill, a loan bill, and a bill to raise more reve-

---

of Congress; R. B. Rhett, Letter to His Constituents, July 4, 1839, in Pendleton *Messenger*, August 30, 1839; White, *Rhett*, 44–45. The secretary of Treasury in January, 1841, suggested a plan that assumed the unalterability of the principles of the Compromise Act, that is, a 20 percent ceiling, discrimination below that if the revenue required it, home valuation, and cash duties. This was not acceptable to Calhoun, and no action was taken in the lame duck session.

53. Malcolm R. Eiselen, *The Rise of Pennsylvania Protectionism* (Philadelphia, 1932), Chap. 7. Interestingly, President William McKinley later said, in his history of the tariff, that the Compromise Act was the principal cause of the panic and depression. See Colton (ed.), *Works*, X, 10–11.

54. Richardson (ed.), *Messages and Papers*, III, 1900–1901.

nue. The last two were made necessary by a huge deficit. He proposed to collect several million dollars additional revenue by levies on the unprotected articles, currently admitted free or charged under 20 percent. This was within the limits of the compromise, in Clay's view. Calhoun, though he had previously concurred, objected that such increases were authorized only after June 30, 1842. The bill sailed through Congress without further controversy, however.[55]

Distribution proved more difficult. Clay took inordinate pride in this measure of his own authorship. It had become a kind of political panacea, not only a solution to the entangled problems of tariff, revenue, and surplus, but the means of financing internal improvements, securing the national domain, and strengthening the Union by raising the stake every state had in it far into the future. "Age after age may roll away," he rhapsodized, "state after state arise, generation succeeding generation, and still the fund [from the proceeds of the public lands] will remain not only inexhaustible but improved and increasing, for the benefit of our children's children, to the remotest posterity."[56]

Unfortunately, this philosophical vision had nothing to do with western realities. Calhoun continued his long-standing opposition to distribution, despite the parallel of the Deposit Act. It was "rank agrarianism"; it was a scheme to revive the tariff and, in effect, to assume the state debts; and it would lead, not to further consolidation, as he had once argued, but to dissolution of the Union—the inevitable result of dispersing the revenue.[57] He insisted, further, that every dollar distributed would increase the danger of breaking the Compromise Act. Clay denied this would be the result, in part because of the effects of home valuation on the customs revenue. He went on to say that the Compromise Act *assumed* distribution of the proceeds of the public lands and the support of the federal government by tariff revenue alone. Cal-

55. *Congressional Globe*, 27th Cong., 1st Sess. (Special), 399–400 and *passim*.
56. Mallory (ed.), *Speeches*, II, 481.
57. Crallé (ed.), *Works*, III, 407–39, IV, 13–43.

houn called this assertion "extraordinary," "pure fiction," unheard of until now. Three senators who had voted for the Compromise Act backed Calhoun; they had never heard of any such assumption or understanding.[58] They were right, at least on the record of legislative debate and discussion in 1833. Clay did not press the point. Nevertheless, lands and tariffs had been firmly linked in public policy deliberations for several years before the compromise; the simultaneous passage of Clay's Land Bill (as it was then called) tended to reenforce the linkage. Indeed, some observers, it may be recalled, had immediately criticized Jackson's Land Bill veto as a violation of the compromise in its cradle.

In 1841 there were two theories of the symbiotic relationship between these measures among the Whigs themselves. Tyler expressed the view more or less common to southern Whigs: if the tariff broke through the ceiling of 20 percent established by the compromise, distribution must be surrendered. Northern Whigs generally held that distribution was necessary to ensure the moderate protectionism allowed by the compromise. This was the position of John Quincy Adams, of Andrew Stewart ("Tariff Andy") of Pennsylvania, and of Clay's erstwhile colleague John M. Clayton. The latter, now in retirement, having appointed himself an authority on the compromise, repeatedly declared that he and his middle states' friends had regarded the distribution bill *as part and parcel of one great revenue and financial system.* Clay had understood their deep interest in this measure to encourage internal improvements in the states and to safeguard the protective tariff, Clayton said, and "offered us the land bill as part and parcel of his grand scheme of protection and compromise."[59] This was, if not "pure fiction," a piece of retrospective symmetry that had little basis in the events making the Compromise Act but stemmed from the controversy over ending it. By logrolling the distribution and bankruptcy bills, Clay got both enacted in

58. *Congressional Globe*, 27th Cong., 1st Sess., 313–15.
59. Speech at Wilmington, Delaware, June 15, 1844, in *Niles' Weekly Register*, August 3, 1844. See also Clayton to Nicholas Biddle [April, 1841?], in John M. Clayton Papers, Library of Congress.

1841, though a proviso had been tacked onto the former suspending its operation if the tariff rose above 20 percent after
June 30, 1842. The long-sought legislation might very well
prove useless, if not because of the proviso, then because
plummeting land sales furnished little money to distribute.

As it happened, the drastic terminal reductions under the
Compromise Act occurred in 1842 when the economy was still
badly depressed. Manufacturing interests, feeling the pinch,
called for a return to protectionism. (Other interests, Louisiana sugar, for instance, joined the call. The planters, who
had acquiesced in the tariff compromise, knowing it must in-
jure them, now considered it as destructive as a pestilence in
their industry.) A new propagandist association, the Home
League, was formed. The ideology that had earlier centered
on the encouragement of domestic industry and the home
market now featured the protection of "high wage American
labor," which, it was said, the tariff had made possible and
which was responsible for the peace, good order, and enlightened conservatism of American society. The cause of col-
lapsing prices, of an epidemic of bankruptcies, of mass un-
employment, and business stagnation was very simple, said
Abbott Lawrence. "We import too much and manufacture too
little." It was one of the unfortunate legacies of the Compromise Act.[60] Fortunately for these interests, however, the deficits of the government made a return to protection, once a
matter of choice, virtually a matter of necessity. When Congress met again, the administration projected a deficit of four-
teen million dollars in the coming year, on top of ten million
in 1841. Moreover, the government had failed to raise the loan
of twelve million authorized in the special session. Tyler reluctantly, therefore, proposed to break the compromise, raise
some duties above 20 percent, and practice protection as an

60. See, for example, the Home League Address, *Niles' Weekly Register*,
March 5, 1842. Lawrence is quoted in the Boston *Courier*, April 21, 1842. On
sugar, see Joseph G. Tregle, "Louisiana and the Tariff, 1816–1842," *Louisiana
Historical Quarterly*, XXV (1942), Chap. 3.

incident to raising revenue. This would trigger the proviso of the Distribution Act and suspend its operation.[61]

Tyler, of course, had lost all influence with the Whigs by his vetoes of successive bank bills in the special session. Clay was the leader—some said the dictator—of the party from his seat in the Senate. He had repeatedly vowed to secure the necessary readjustment of the tariff before going into a long-deferred retirement. Everyone understood the move was preparatory to his presidential candidacy, in which he hoped to make the tariff the dominant issue. A decade earlier the tariff had been primarily a sectional issue; now, though it had not entirely lost that character, it was primarily an issue between the two great political parties. This was as Clay wanted it. In February he introduced a series of resolutions setting forth "a system of policy" for the Whigs. Its principal measure was a new tariff. The rate that was to fall to 20 percent on July 1 would be raised to 30 percent; and home valuation would add another 5 percent. The revenue would thus be increased by an estimated six or seven million dollars. With returning prosperity, this ought to be sufficient for the needs of the government as well as for the protection of manufactures. But Clay emphasized that revenue, not protection, was the ruling principle. "There is no necessity of protection, for protection," he declared cryptically. The resolutions also called for repeal of the proviso of the Distribution Act.[62] In the ensuing senatorial debate Clay and Calhoun again exchanged recriminations on the Compromise Act. Nothing he now proposed violated the spirit or the letter of that act, Clay insisted. The sophistry of the argument, deducing protectionism from the compromise, amazed Calhoun. "There is no estimating the force of self-delusion in a position so contradictory," he declared.[63] Webster doubtless agreed. From his post as secretary

61. See especially Tyler's special message, March 25, 1842, in Richardson (ed.), *Messages and Papers*, III, 1961–62.

62. *Congressional Globe*, 27th Cong., 3rd Sess., 235–36, 156; Mallory (ed.), *Speeches*, II, 532–61.

63. Crallé (ed.), *Works*, IV, 109.

of state he advocated a forthright return to protectionism, abandoning the "horizontal" principle of a uniform *ad valorem* tariff in favor of specific duties affording outright, not incidental, protection, and giving up home valuation as impracticable, if not unconstitutional as Calhoun maintained.[64]

Clay retired at the end of March, before the committees of either house reported their tariff bills. When it became apparent that no new law could be enacted before the expiration of the old one, Congress voted to extend the present rate of duties one month, thereby blocking the final reduction to 20 percent under the Compromise Act, and provided, further, that this should not suspend the Distribution Act. But Tyler, invoking the sanctity of the compromise, vetoed the bill. When July 1 arrived, the ultimate goal of the 1833 act—the hypothetical revenue standard—was realized. The administration adopted Clay's position on the continuing force of the statute; and when this was challenged, it was upheld by the Supreme Court.[65] Early in August Congress succeeded in passing a new tariff bill. While Clay's resolutions had maintained at least the semblance of fidelity to the compromise, the bill was unashamedly protectionist. It returned the tariff generally to the 1832 scale, with most protective duties under 35 percent, though the bill also made generous use of minimums and specific duties on woolens, iron, and several other manufactured articles as well as on selected commodities like sugar. Home valuation was abandoned. The proviso of the Distribution Act was repealed—an incident in the political vendetta between the Whigs and Tyler. Although the vote in both houses reflected the intensely partisan division on the tariff, a substantial number of southern Whigs defected in protest against the repeal, which they considered a breach of faith. It was on this ground, and this ground only, that Tyler vetoed the tariff bill.[66] Predictably, he adhered to his position that if the tariff went up, distribution must go down. Many

64. "Draft for Annual Message," December, 1841, *Writings and Speeches of Daniel Webster,* XV, 140–43.
65. *Aldridge et al.* v. *Williams,* III Howard (U.S.), 9–32.
66. Richardson (ed.), *Messages and Papers,* III, 2034–36.

northern Whigs, at this point, sullenly opposed sacrificing a leading party measure, distribution, as the price for a tariff and revenue bill. But the deed was done, and thus ended the incestuous union between tariff and distribution.

In the South, and among Democrats everywhere, the new tariff was widely regarded as a betrayal of the compromise. To Calhoun's followers in South Carolina, particularly, it seemed that the state had returned full circle to 1832, burdened by the same tariff, cursed by the same economic conditions. Naturally, the same remedy suggested itself. But in 1842 Calhoun was running hard for the Democratic presidential nomination and kept his friends on a tight leash.[67] Clay, at Ashland, had no direct responsibility for the new tariff. Since he was courting southern support for his presidential candidacy, there were hopes in that quarter that he would repudiate the act and call for return to the compromise with which his fame and his politics were so closely identified. But Clay, after some initial doubts, acquiesced in the act; before long he engraved it on his banner.[68]

George McDuffie, one of those who had hoped for greater fidelity from Clay, came out of political retirement in order to overthrow the new tariff. Upon his return to Congress in December, 1843, he introduced a bill to restore the 20 percent ceiling of the Compromise Act in two years. He was full of praise for the author of that act. "I then said—I have always said—that never in the course of my political experience, had I known any public man display a more heroic moral courage than the Senator did on that memorable occasion."[69] Distressingly, Clay now approved the destruction of his best work.

67. In retrospect, James H. Hammond thought that South Carolina should have promptly nullified the tariff of 1842—indeed that this was an obligation entailed by the repeal ordinance of March, 1833—but that the state did not because of Calhoun's political ambitions. See his Diary, October 25, 1844, in James H. Hammond Papers [Microfilm], Library of Congress.

68. "The Tariff," *Southern Quarterly Review*, II (1842), 428–34; Washington, D.C., *Madisonian*, July 18, 1844; Clay to J. M. Berrien, August 15 and September 4, 1842, in Typescripts, Clay Papers.

69. *Congressional Globe*, 28th Cong., 1st Sess., 105. See also the report of McDuffie's speech in *Enquirer*, July 8, 1844.

McDuffie's effort met with no success in the Senate, nor did anything come of a similar measure in the House despite an overwhelming Democratic majority. Obviously, northern Democrats had no taste for the tariff as an issue in the impending election, in which they expected Van Buren would be matched against Clay.

Another issue, the annexation of Texas, even less welcome than the tariff to northern Democrats, determined the nomination of James K. Polk over Van Buren and then cast a shadow over the presidential campaign. The tariff remained in the foreground of controversy, however. In South Carolina, where for a decade the Compromise Act had been considered a treaty between belligerent parties, Robert Barnwell Rhett led a revolt against Calhoun's quiescent leadership, demanding a state convention for the purpose of nullifying the tariff or seceding from the Union. Calhoun regained control in time to keep the Palmetto State in the Democratic fold, casting its electoral vote for Polk, from whom he had obtained private assurances of tariff reform. In an outrageous fraud on the electorate of the northern states, Democrats advertised Polk as a greater protectionist than Clay, the Father of the American System. Moreover, this Janus-faced appeal to sectional prejudice was projected onto Clay. Democratic doggerel portrayed the Kentuckian as a dissembler and a hypocrite on the tariff.

> Orator Clay had *two tones* in his voice;
>   The one squeaking *thus*, and the other down *so*;
> And *mighty* convenient he found them both—
>   The squeak at the *top* and the guttural *below*.
>
> Orator Clay looked up to the North;
>   "I'm for the tariff PROTECTIVE," said he;
> But he turned to the South with *his other tone*!
>   "A tariff for revenue only 't will be!"[70]

Clay was not blameless, of course. He had attempted to ride the Compromise Act in opposite directions; miraculously, as it turned out, protectionism and "revenue only" coexisted

---

70. *Congressional Globe*, 28th Cong., 1st Sess., 662.

much better than had seemed possible in 1833. He could hardly be faulted for playing up the moderation of the compromise
in the South, where he made a great tour in the spring. Before
huge crowds in Savannah, Milledgeville, Charleston, and other southern cities he flatly rejected Calhoun's conception of
the compromise and asserted, as he had repeatedly, its true
principle to be discrimination within the revenue standard.
He glossed over the differences between that principle and
the 1842 act but said, in mitigation, "If violation had occurred
on one side, it had also occurred on the other," as in the abandonment of home valuation.[71] The growth of moderate protectionist sentiment in the South, brought on by the planters'
disillusionment with the trumpeted benefits of "free trade,"
by increasing industrialization, by the influence of the Whig
party, had inspired Clay with false hopes, as the result would
show, of collecting a political debt long overdue.

In the North the Whig standard-bearer may have lost the
key protectionist state of Pennsylvania because of Polk's ruse
and long memories of the Compromise Act. The idea that he
had betrayed the American System—"murdered his own
child"—continued to haunt him. Webster had never forgiven
him; and their relations had become embittered when Webster
remained in his cabinet post after the Tyler administration was
disavowed by the Whig party. At the time of the tariff debate
some months later, Webster transmitted to Henry A. Wise,
a Virginia congressman and one of the "corporals' guard"
around Tyler, a statement intended to embarrass Clay on his
banner issue. According to this statement, given on the authority of "one who saw the manuscript draught of the act of
1833 before it was offered," the clause (section 3 in the act)
providing that from June 30, 1842, "duties shall be laid for the
purpose of raising such revenue as may be necessary to an
economical administration of the Government," was followed
in the draft by these additional words: "And such duties shall
be laid without reference to the protection of any domestic

71. Speeches reported in *Niles' Weekly Register*, April 20, 1844, and *National
Intelligencer*, April 16, 1844. See also the speech of Richard Yeadon, Charleston,
in defense of the Compromise Act, in Charleston *Courier*, October 24, 1844.

articles whatever." Wise inserted Webster's pernicious paragraph into a public letter to his constituents after Congress
adjourned. It was part of the "secret history" of the compromise, he said, and revealed its true purpose.[72] The Polk campaign's subsequent portrayal of Clay as an enemy of protection stemmed, in part, from this revelation. Political insiders
had no difficulty identifying Wise's unnamed informant. "Our
modern Anthony, it seems, means no longer to succumb,"
the *Globe* editorialized. "Here, then, for the first time, Mr.
Webster displays the courage of a man."[73] However that may
be, it was not until the 1844 compaign that Webster was revealed as the author, unwillingly, by the Tylerites in retalia-
tion for his return to the Whig party.[74]

The attack on Clay's protectionist credentials drew from his
old colleague Clayton a full defense of the Compromise Act,
founded on its "secret history," which was widely circulated
during the campaign. He refuted "the common error" that
the act looked to "a horizontal tariff of 20 percent." It was intended, rather, to secure moderate protectionism under the
revenue standard. The 1842 tariff was in substantial compli-
ance with the pledge of the compromise, Clayton asserted, except for the elimination of distribution and home valuation.[75]
Clay endorsed this account and added several details from his
own recollection. He agreed with Clayton that the great triumph of the compromise was in placing the protective policy

72. The letter is in the *Madisonian*, September 24, 1842. The words quoted
from the alleged manuscript draft vary somewhat from those contained in
the manuscript of the "reported draft," endorsed as "Mr. Clay's first project,"
in the Huntington Library. But it seems likely that Webster had reference to
this document, or one like it, probably a personal copy he had made from the
original, and also that that original was, indeed, "Mr. Clay's first project," the
one he had formed in Philadelphia and handed around in Congress early in
January, 1833.

73. *Globe*, October 3, 1842.

74. *Madisonian*, August 12, 14, 1844; Boston *Courier*, August 17, 1844.

75. In his Wilmington speech, *Niles' Weekly Register*, August 3, 1844. See
also the New York *Tribune*, June 18, 20, 1844, which announced a large pam-
phlet edition of the speech.

out of reach of Jackson's violence. And he recalled saying to his protectionist adversaries at the time, "Let us take care of ourselves now; the people of 1842 may be trusted to take care of themselves. Public opinion, in the meantime, may become more enlightened, and the wisdom of the protective policy may be demonstrated." He had not been disappointed. "Everywhere the cry is for a tariff for revenue, with discriminations for protection. Everywhere the preservation of the tariff of 1842 . . . is loudly demanded." As to the accusation of having abandoned protectionism, Clay said it would distress him exceedingly if he were not already accused of every crime in the Decalogue. Clayton published the letter as part of another speech in this continual intercourse of history with politics.[76] Clay, ever the pragmatist, cared little for the historical part. "Is it not somewhat of a collateral if not obsolete issue?" he nudged Clayton.[77] When he read in the Tyler newspaper, the *Madisonian*, the paragraph attributed to Webster he supposed it was in error, but even if the original draft contained a clause foredooming protection, it was not in the final bill, which alone mattered.[78]

The public history of the Compromise Act ended, whimperingly, with the Democratic "Walker Tariff" of 1846. Based upon the elaborate report of Polk's secretary of the Treasury, Robert J. Walker, it adopted the rule of maximizing the revenue regardless of protectionist considerations, and it further determined that a uniform 20 percent *ad valorem* rate of duties would yield the maximum revenue. The correlation of high revenue with low to moderate duties was not supported by the American experience; it was, in fact, contradicted by the overflowing Treasury under the 1842 act; but the Walker plan, with some modifications, became law on a straight party-line

76. Clay to John M. Clayton, August 22, 1844, in a speech at Lancaster, Pennsylvania, in *Niles' Weekly Register*, September 14, 1844. The letter, slightly abridged, is in Calvin Colton, *Life and Times of Henry Clay* (New York, 1846), II, 259–61.
77. Clay to Clayton, N.d., [September (?), 1844], in Clayton Papers.
78. *Ibid.*, August 29, 1844.

vote. Even so, Democrats had the audacity to declare that the act was the legitimate fulfillment of the compromise, one that Clay might have recognized as his own had he not weakened and given in to high protectionist interests in 1842.[79] Enacted only a few weeks after Parliament repealed the Corn Laws, opening the British market to American agricultural productions, it was thought by keen observers then, and by some historians since, to have been a fair exchange for that repeal, perhaps with the settlement of the Oregon boundary thrown in.[80] The act was, at any rate, consonant with a movement in the Atlantic world toward freer trade. It endured for eleven years, as long as the country rode the wave of prosperity that commenced in 1844. New questions, most of them centered around slavery, only a cloud on the horizon at the time of the nullification crisis, dominated the nation's political life. The Great Compromiser, in 1850, labored to make a compromise of these questions, too; and as long as men struggled to preserve the Union they would remember the Compromise of 1833, not alone for its political genius but for its instruction in the perils and perplexities of compromise in the American republic.

A complex event like the Compromise of 1833 may in the final analysis be considered historically important for either or both of two reasons. First, it was a crucial causative influence in a course of events big with the fate of the nation. Thus it has been said that the compromise "occasioned the great financial crisis of 1837" or, more significantly, that it was the "great misfortune . . . which led the country finally into civil war."[81] The workings of history are much too subtle and intri-

79. For example, Robert Dale Owen, in *Congressional Globe*, 29th Cong., 1st Sess., 1004–1005; Garrett Davis to the Editors, in *National Intelligencer*, August 10, 1844; *ibid.*, August 18, 1844.

80. For example, New York *Journal of Commerce*, quoted in *Niles' Weekly Register*, February 14, 1846.

81. McKinley, in Colton (ed.), *Works*, X, 10–11; John W. Burgess, *The Middle Period, 1817–1858* (New York, 1897), 239. See also Hermann von Holst, *Constitutional and Political History of the United States* (New York, 1876), I, 505. "It was a terrible victory [for South Carolina]; the vanquished have been terri-

cate for such stupendously simple explanations, however. Whatever may have been the main causes of the Panic of 1837—a matter on which historians disagree—the Tariff of 1833 was surely not one of them, though it may have retarded recovery from the ensuing depression. This belief, as we have seen, sparked the revival of protectionism and in 1842 resolved the fundamental ambiguity of Clay's legislation on the side of protection rather than of revenue. As to the Civil War, the compromise, it used to be argued, amounted to a virtual acknowledgment of the right and power of nullification, which plunged the nation on the downward course that led to secession and war twenty-eight years later. But, of course, the argument ignored the equivocal character of the compromise. Even if Clay's tariff could be interpreted as a surrender, over his vehement denial, Jackson's proclamation and the Force Bill provided "the great education in national principles" that, in the opinion of other historians, rallied the northern people for the preservation of the Union, one and indivisible, in 1861.[82] In truth, both interpretations mistake the symbolism of the crisis over nullification for historical cause and effect. In that regard, all that can be said with perfect assurance is, first, that the compromise ended the immediate crisis provoked by nullification, and second, that it settled for almost a decade the politically explosive issue of the protective tariff. These were important historical effects in themselves; but it is apparent, from the perspective of 1842 or of 1861, that they did not alter the fate of the nation.

The greater import of the compromise for the historian lies not with momentous questions of causation but with the description of a range of phenomena that help to explain the political process and also add texture and tone to our understanding of the past. Let me offer several observations, by way of general conclusions, in this area. First, the process of policy making in the republic of Jackson and Clay was in-

---

bly scourged for the defeat suffered through their sin and the victors have been scattered to pieces."
82. Burgess, *Middle Period*, 241.

credibly complex. It involved a gathering of forces, some converging, some diverging, none readily submissive to common goals or direction: the traditions of the Constitution and of Jeffersonian republicanism; the clash of sections and economic interests; the ambitions of politicians and the stratagems of parties and factions; the rivalry between the president and Congress and the character of leadership in these two great branches of government; the ideals of nationalism and states' rights, with the corresponding fears of disintegration and consolidation; and the intricate relationships among policies that drew upon diverse or opposing clienteles. All these factors and forces entered into the making of the Compromise of 1833. It was, as we have seen, the product not of a simple bargain between opposing sides but of a whole complex of trade offs among leaders, clients, and interests. Second, the political life of the time, from the rhetoric of debate to the decision on specific issues, was dominated by a few leaders who enjoyed the same celebrity and power on the public stage as the great dramatic actors of the age enjoyed in the theater. The history of the Compromise of 1833 is largely the history of Clay and Calhoun, Jackson and Webster; it cannot be explained without them. Not only did the dominant leaders become personifications of ideas, issues, and policies; it often seemed, as with Clay's tariff bill, that policy was finally determined by little more than personal political whim, passion, or ambition. This factor of personality marks the most striking contrast to the process of policy making in our time, when impersonal bureaucratic and institutional factors loom so much larger. Finally, the Compromise of 1833 vindicated the arts of compromise in American politics. To be sure, it did not permanently solve anything. The problem of the surplus remained; the conflict over the tariff was suspended, unresolved, in a state of ambiguity; and the Union, while preserved, was still vulnerable. However, it is not the purpose of political compromise to *solve* great issues but rather, by mediating between the methods of persuasion and of force, to find a proximate solution, one that will practically overcome the

immediate difficulty, restore order and harmony, and enable
the society to get on with its business. In accomplishing this,
the Compromise of 1833 justified the faith of its chief architect
in the workings of concession and conciliation as the price of
Union under the United States Constitution.

Index